W9-AUV-221

Zoology

Disclaimer

Adult supervision is required when working on these projects. No responsibility is implied or taken for anyone who sustains injuries as a result of using the materials or ideas, or performing the procedures put forth in this book. Use proper equipment (gloves, protective clothing, safety glasses, etc.), and take other safety precautions, such as tying up loose hair and clothing and washing your hands when you finish working.

Use chemicals, dry ice, boiling water, or any heating elements with extra care. Hazardous chemicals and live cultures (organisms) must be handled and disposed of according to appropriate directions set forth by your sponsor. Follow your science fair's rules and regulations and/or standard scientific practices and procedures as set forth by your school or other governing body.

Additional safety precautions and warnings are mentioned throughout the text. If you use common sense and make safety a first consideration, you will create a safe, fun, educational, and rewarding project.

Zoology
High School Science Fair Experiments

H. Steven Dashefsky

TAB Books
Division of McGraw-Hill, Inc.
New York San Francisco Washington, D.C. Auckland Bogotá
Caracas Lisbon London Madrid Mexico City Milan
Montreal New Delhi San Juan Singapore
Sydney Tokyo Toronto

Published by TAB Books, a division of McGraw-Hill, Inc.

pbk 1 2 3 4 5 6 7 8 9 0 DOC/DOC 9 9 8 7 6 5 4
hc 1 2 3 4 5 6 7 8 9 0 DOC/DOC 9 9 8 7 6 5 4

Library of Congress Cataloging-in-Publication Data
Dashefsky, H. Steve.
Zoology : high-school science fair experiments / by H. Steven Dashefsky.
p. cm.
Includes bibliographical references and index.
ISBN 0-07-015686-7 ISBN 0-07-015687-5 (pbk.)
1. Zoology projects. 2. Zoology—Experiments. I. Title.
QL52.6.D375 1994
591'.076—dc20 94-29631
CIP

Acquisitions editor: Kim Tabor
Editorial team: Joanne Slike, Executive Editor
David M. McCandless, Supervising Editor
Karen Collier, Book Editor
Production team: Katherine G. Brown, Director
Rhonda E. Baker, Coding
Brenda S. Wilhide, Computer Artist
Rose McFarland, Desktop Operator
Lorie L. White, Proofreading
Joann Woy, Indexer
Designer: Jaclyn J. Boone

GEN1
0156875

Contents

Appendices

Introduction

How to use this book

There are two ways to use this book. If you are new to science fair projects and feel you need a great deal of technical guidance, you can use the projects as explained in this book with few if any adjustments. These are good, solid science fair projects. However, if you want to be a contender with an award-winning project, you must use the experiments in this book as only the starting point. Look for ways to modify, enhance, or otherwise customize to make it your project. Suggestions on how to do this are given in the Going Further and Suggested Research sections of each project.

Each project in this book covers the following topics:

- Background
- Project overview
- Materials list
- Procedures
- Analysis
- Going further
- Suggested research

Background

This section provides background information about the topic to be investigated. It offers you a frame of reference, so you can see the importance of the topic and why research is necessary to advance our understanding. This section can be considered the initial step in your literature search. Although a small step, it is enough to see if the subject peaks your interest. (See Getting started for more on your literature search.)

Project overview

If the background section interests you, continue reading because the project overview section describes the purpose of the project. It explains the problems that exist and poses questions that the experiment is intended to resolve. These questions can be used to formulate your hypothesis. Be sure to discuss this section, as well as the next, with your sponsor to see if the requirements imposed can realistically be met.

Materials list

This section gives a list of everything needed to perform the experiment. Be sure you have access to or can get everything before beginning. Some equipment or apparatus are expensive. Check with your teacher to see if all the equipment is available in your school or if it can be borrowed. Be sure your budget can handle everything that must be purchased. A list of scientific supply houses is provided at the back of the book.

Although most people don't think of research scientists as a group of handy men and women, they must be. Building a device or experimental workstation often involves many trips to the hardware store for supplies, a little sweat on the brow, and a lot of ingenuity.

Living organisms, such as microbial cultures, probably will have to be ordered from scientific supply houses. Others, such as insects, can be either ordered, purchased locally, or caught, depending on the project, your location, and the time of year. If you are using live organisms, work with your sponsor to be sure you adhere to all science fair regulations and standard biological research practices. Before beginning, discuss with your sponsor the proper way to dispose of any hazardous materials, chemicals, or cultures. If you plan to use people as subjects in your project, be sure to read the section below, a special note about human subjects.

Procedures

This section gives step-by-step instructions on how to perform the experiment and suggestions on how to collect data. Be sure to read through this section with your sponsor before undertaking the project. Illustrations are often used to clarify procedures. Although each step is given, some projects require standard procedures such as inoculating a petri dish. These steps are often stated but not explained. Your sponsor can help you with these standard procedures.

Analysis

This section doesn't draw conclusions for you. Instead, it asks questions to help you analyze and interpret the data, so you can come to your own conclusions. Many projects contain empty tables or charts to help you collect and analyze the data. You should convert as much of your raw data as possible into line and bar graphs and/or pie charts. In some cases these also are included in the project.

Consider using a computer to speed-up your data collection and analysis and to enhance your report. If you have access to a personal computer with a graphics program, such as Harvard Graphics or MacDraw, it will make drawing graphs, tables, and charts much easier. Spreadsheet programs, such as Lotus 1-2-3 and Microsoft Excel, can generate graphs directly from your raw data. You simply enter the numerical data collected into the spreadsheet and then instruct the program to generate the graph. (See Fig. I-1.)

EXPERIMENTS				
GROUP	A	B	C	D
1	~	~	~	~
2	~	~	~	~
3	~	~	~	~
CONTROL	~	~	~	~

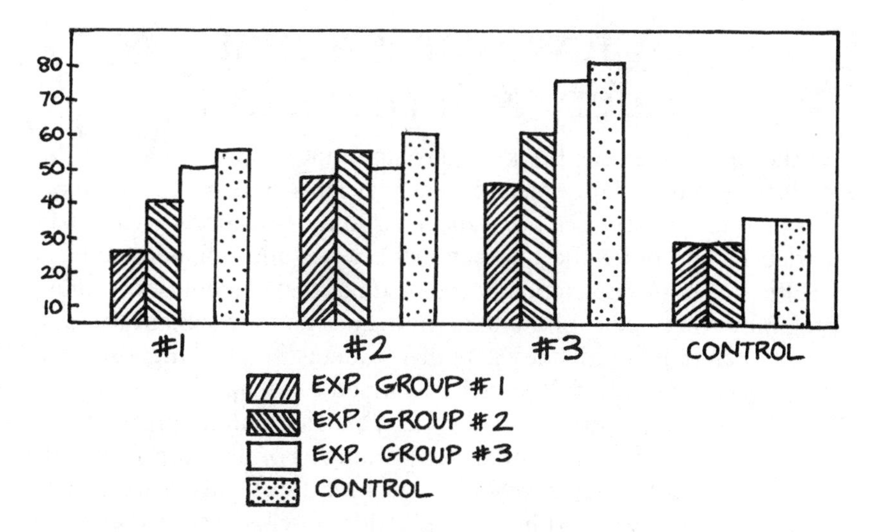

I-1 *An excellent way to analyze data is to use a spreadsheet software program capable of creating tables and graphs from your raw data.*

Some experiments might require statistical analysis to determine if there are significant differences between the experimental groups and the control group. Check with your sponsor to see if you should perform statistical analyses for your project and if so, what kind.

Going further

This section is an important part of each project. It lists ways you can continue researching the topic beyond the original experiment. These suggestions can be followed as is or, even more importantly, they can be used to spark your imagination to think of some new twist or angle to take while performing the project. These suggestions might show ways to more thoroughly cover the subject matter and/or show you how to broaden the scope of the project.

The best way to ensure an interesting and fully developed project is to include one or more of the suggestions from the going further section or include an idea of your own that was inspired from this section.

Suggested research

This section suggests new directions to follow while researching the project. It often suggests what to read and organizations, companies, or other sources to contact. Using these additional resources could turn your project into a winner.

A few words about safety & supervision

All the projects in this book require an adult sponsor to assure the student's safety and the safety of others. Science Service, Inc. is an organization that sets science fair rules and regulations and safety guidelines and holds the International Science and Engineering Fairs (ISEF). This book recommends and assumes that students performing projects in this book follow ISEF guidelines as they pertain to adult supervision. ISEF guidelines state that students undertaking a science fair project have an Adult Sponsor assigned to them.

The Adult Sponsor is described as a teacher, parent, professor, or scientist in whose lab the student is working. (For the purpose of this book, this will usually be the student's teacher.) This person must have a solid background in science and be in close contact with the student throughout the project. The Adult Sponsor is responsible for the safety of the student while conducting the research, including the handling of all equipment, chemicals, and organisms. The sponsor must also be familiar with regulations and commonly approved practices that govern chemical and equipment usage, experimental techniques, the use of laboratory animals, cultures, and microorganisms, and proper disposal techniques.

If the adult sponsor is not qualified to handle all of these responsibilities, the sponsor must assign other adults who can fulfill these responsibilities. Most science fairs require appropriate forms that identify the Adult Sponsor and his or her qualifications to be filled out before proceeding with a project.

The sponsor is responsible for reviewing the student's research plan, as described later in this book, and making sure the experimentation is done within local, federal, and ISEF (or other appropriate governing body's) guidelines.

The entire project should be read and reviewed by the student and the Adult Sponsor before beginning. The Adult Sponsor should determine which portions of the experiment the student can perform without supervision and which portions will require supervision.

For a copy of the ISEF's rules and regulations, contact

Science Service, Inc.
1719 N. Street, N.W.
Washington, DC 20036
(202) 785-2255

The booklet includes a checklist for the Adult Sponsor, approval forms, and valuable information on all aspects of participating in a science fair.

Special thanks

Many of the experiments in this book were adapted from original International Science and Engineering Fair (ISEF) projects. I want to thank the following young scientists for their outstanding projects and wish them the best of luck in their future scientific endeavors. (In most cases, the projects were modified and edited for this book.)

- Paul Lynch for "Video games & your heart: Is there a relationship between video games & heart stress?" (chapter 4)
- Ian Robert Lemieux for "Age & smell: Does aging affect your sense of smell?" (chapter 5)
- Abby Janoff for "Sight deprivation & hearing: Does short-term loss of sight enhance your ability to hear?" (chapter 6)
- Stephen DeCost for "Music & the body: Do different types of music affect our bodies differently?" (chapter 7)

- Benjamin Martin for "Natural pesticides: Do plant extracts contain any natural pesticides?" (chapter 8)
- Diana Murphree for "Slug prevention: Can a mechanical barrier keep slugs out of your garden?" (chapter 9)
- John Howard for "Herbs & germs: Can herbs inhibit the growth of microbes & the spread of disease?" (chapter 11)
- Kristen Baker for "Slug attractants: What substances can attract & kill slugs?" (chapter 12)
- Christian Marcillo for "Madagascar hissing cockroaches: Are all their hisses alike?" (chapter 13)
- Megan Duncan-Smith for "Color change in anoles (American chameleons): Does the lizard's size affect the speed with which it changes color?" (chapter 14)
- Stacey Favaloro for "Vitamins & regeneration: Do vitamins affect the regeneration of planaria?" (chapter 17)
- Kelly Lincicum for "Second-hand smoke: Does second-hand smoke affect mealworms?" (chapter 18)
- Tobey Kay Cho for "Growth regulator hormones: Do these new herbicides affect the regeneration of planaria?" (chapter 19)
- Sarah E. Lafaver for "Heavy metal pollution: Does copper sulfate pollution harm brine shrimp?" (chapter 20)
- Alicia Gomes for "Fossil fuels & animals: Does coal fly ash harm mealworms?" (chapter 21)

Part 1

Before you begin

Before delving into any science experiment, you need to understand three things: the terminology used, the methodology required, and the suitability of the experiment to your own situation. The following three chapters examine these elements.

1

An introduction to zoology

Zoology is the science of animal life. It is a very broad topic that includes animals of all sizes and shapes, ranging from organisms invisible to the naked eye, such as an amoeba, to enormous specimens that dwarf our species, such as the 100-foot-long blue whale. Because zoology covers such a wide range of animals, it is divided into subdisciplines. For example, invertebrate zoology is the study of animals without backbones, while vertebrate zoology is the study of those animals with backbones. It can be divided into smaller groups, such as entomology, which is the study of insects, and mammalogy, which is the study of mammals.

Roughly one and a half million types of organisms (plants and animals) have been identified on our planet with about one million of them being animals. Animals are found almost everywhere on our planet. They can be found on and in the land, from the elephant stomping across the savanna to a community of millions of microscopic nematodes living in the topsoil.

Animals are found in all but the most polluted bodies of water, from zooplankton in ponds to mosquitoes that mature in puddles to large predatory fish and mammals that inhabit the open oceans. Birds and bats fly through the air, while some invertebrates catch rides on wind currents. Animals feed on plants and on each other. Many are parasites that feed on other animals, leaving them unharmed, while others are parasitoids, sapping the life out of their hosts. Still others find their food in the dead, decaying bodies of plants and animals.

Even then William was building twenty-five homes along the Ring of Kerry that would be worth millions. Certainly nobody within the Tigers organization, including players, would have objected to Teahan getting gas money.

Connie and I remained in Tralee for the holidays. My pay cut and her flight had depleted our funds, and with the Tigers doing so much better, I didn't feel an urgency to get out of town. Instead we loaded the freezer up with fresh fish, stocked up on turf and coal for the fireplace, and had a romantic Christmas at home. Also, Kieran Donaghy's mother made us two porter cakes.

The Sword in the Hand

By the end of January the Horan's Health Store Tigers were 11–2, and sailing. We were so far ahead of the second-place Demons of Cork, who were 7–5, that only a major disaster in the final seven games could keep us from being crowned Southern Conference champions of the Irish Super League. That meant an automatic bid to the Final Four. It was hard to believe, dizzying even. People I'd never met in town were calling, "Well done, Coach!" We were getting color photos in the three local papers. The crowd of youngsters at games continued to grow.

Paddy Jones picked out a difficult tune for me to learn that week called "Tom Billy's Reel." It was a tune that was characteristic of Kerry fiddlers—from the playing of a long-gone blind fiddler named Tom Billy Murphy—but more subtle than much of what I'd been learning.

"Can you slow it down for me?" I asked, as we were, rather, I was struggling through.

"Certainly," Paddy said, and we started in again, but slower. Fiddling together with Paddy was the best part of the lesson, and getting entrained with him was becoming less elusive. But that afternoon, trying to slow down "Tom Billy's Reel," Paddy was screwing it up too. We ground to a halt. Could entrainment work both ways? Were my problems and limitations now rubbing off on Paddy?

Paddy set his fiddle on his lap. He knew what the trouble was. "Sometimes playing is like going across an icy pond," Paddy said. "We don't get much ice around here anymore. Ireland has gotten warmer. When I was a boy, the ponds would freeze once or twice a winter. And if you walked across

What separates animals from plants?

Animals are consumers, rather than producers. This means animals must eat (consume) their food, as opposed to producing their own food by using energy from the sun. When an animal consumes its food, it gains nourishment that contains stored, chemical energy. The animal digests this food, which then releases the stored energy. When this energy is released, it is used by the animal to move, grow, maintain its body, and reproduce. In some cases, it is used to produce light, as with the common lightning bug (which is really a beetle).

Plants on the other hand, convert sunlight (radiant energy) directly into chemical energy during the process of photosynthesis. This is the primary difference between plants and animals. Other differences are even more obvious. Animals can generally move about freely, while plants cannot. Almost all animals have a nervous system of some sort that allows them to react rapidly to a stimulus such as an attack from a predator. Plants don't have nervous systems and responses to stimuli occur very slowly.

Another important difference between plants and animals is in the structure of their cells. Animals have a very thin cell membrane around the cells in their bodies. This membrane usually does not provide any kind of support to the organism, and it is permeable, meaning certain materials can pass through it. On the other hand, most plants have a thick rigid cell wall that helps support the organism and does not allow materials to pass through.

The foundations of animal life

Biodiversity refers to the uniqueness found throughout the animal kingdom. In spite of this biodiversity, every animal must perform these five basic functions:

- Ingest and digest food so it can be transported to all the cells in the body.
- Respire, which releases the chemical energy stored in the food.
- Respond to stimuli such as light, heat, or a predator.
- Remove the waste products it produces while performing functions.
- Reproduce to continue the species.

Let's look at each function.

Ingesting, digesting, and transporting food can be as simple as bringing a food particle through the cell membrane, adding a few chemicals and dispersing it throughout the cell, as in an amoeba. Higher forms of life, such as mammals, have complex digestive systems to get the food into the body and into a usable form, and circulatory systems to distribute the nutrients to all the cells throughout the body.

Respiration is the biochemical process of breaking down sugar molecules to release the chemical energy stored within so it can be used by the organism to survive. Animals use the released energy to move, transport substances within their bodies, grow and maintain their body tissues, and reproduce. Some even use it to produce light. Many marine organisms produce this light, called bioluminescence.

Animals must respond to their environment. Whether it is a one-celled animal moving away from light or a hyena attracted to a rotting carcass, some form of nervous system controls their bodies, allowing them to respond appropriately. (See Fig. 1-1.)

1-1
Animals respond to their environment using some form of nervous system.

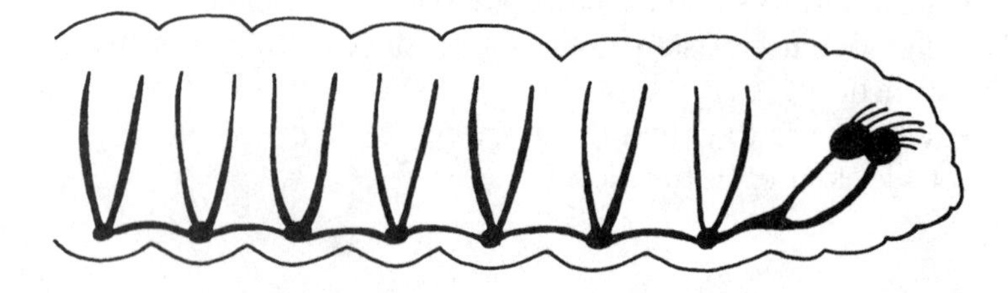

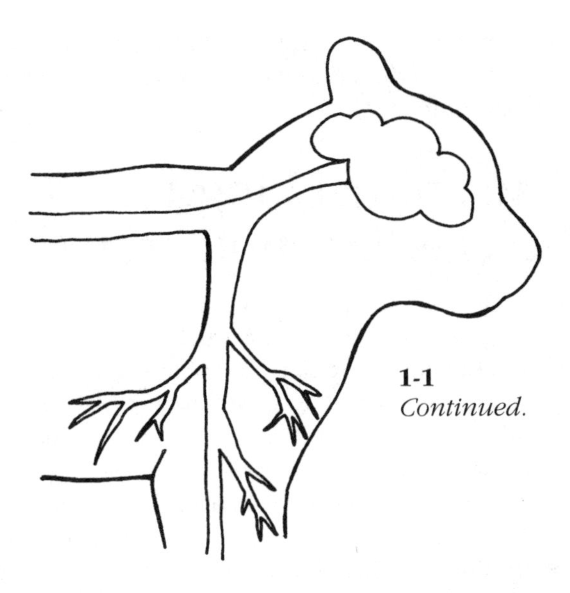

1-1
Continued.

What an animal eats but doesn't use becomes waste. Most biochemical processes produce waste products as well. All this waste cannot be left to accumulate in the animal's body or it would poison itself. Simple, one-celled animals only have to pass wastes out through their cell membranes, but more complicated forms of life require specialized organs. These organs help remove wastes from the cells and from our bodies and include our kidneys and sweat glands.

Finally, animals must reproduce if the species is to survive. Some simple, one-celled animals reproduce by fission, meaning they simply divide into two individuals. Others can create new individuals by budding. Still others produce young asexually without the need of two sexes. Most higher forms of animals, however, reproduce sexually by the joining of egg and sperm.

The stuff of life

What else do people and amoebas have in common besides the five items mentioned above? Water. All animals are made mostly of water, ranging from about 50 to 90 percent. This is because an organism's body consists of cells, and the cells are filled primarily with water. Cells are filled with a fluid called cytoplasm, which is composed of proteins, carbohydrates, fats, salts and, of course, water.

The stuff of life is also the stuff of inanimate (lifeless) objects. The materials found in living cells are also found in the non-living world. These chemicals pass from the rocks, soil, water, and air into living organisms, primarily by the green plants during photosynthesis. These substances are then passed through food chains and food webs to all forms of life, only to be returned to the non-living world when the organisms die and decompose.

The animal kingdom & the classification system

Scientists classify all forms of life according to how closely related they are to one another. You can compare this classification system to a file cabinet that contains many drawers, with each drawer containing many folders and dividers. Each organism is placed on its own sheet of paper that must be placed somewhere in this file cabinet.

The more closely related two types of animals are, the more closely their sheet of paper will be placed in the cabinet. For example, an insect and a dog would be placed in two different drawers since they are so different. However, a dog and a cat would appear not only in the same drawer, but within the same folder, since they are so similar.

To continue this comparison, imagine there are two file cabinets; one is labelled "Plants" and the other "Animals." Each cabinet represents a "kingdom." Let's concentrate on the animal cabinet. Imagine there are ten drawers in this cabinet. Each drawer represents a "phylum" (phyla is plural). Each phylum contains animals that have certain characteristics in common. Insects and spiders are found in one of the drawers labelled, "Arthropods," since they both have a hard skeleton-like shell called an exoskeleton. Dogs and cats are found in another drawer labelled, "Chordata," since they have backbones.

Inside each drawer are a series of folders with labels. These folders represent "classes" of organisms that have even more similar characteristics. For example, our insect and spider both go into the same drawer (phylum) but will go into different folders within the drawer. The insect is placed in a folder labelled, "Insecta," which contains all the animals with six legs. The spider, however, is placed a folder labelled, "Arachnida," since it has eight legs.

Our dog and cat are both placed in a drawer (phylum) labeled "Chordata." They not only go in the same drawer, but are also placed in the same folder (class) labelled, "Mammalia," since they both have fur and mammary glands. Since the dog and the cat are in the same folder, they must be more closely related to each other than the insect and spider, which are in the same drawer but different folders.

The classification system continues to distinguish which animals are most closely related by placing them in "orders," "families," "genera," and finally "species." Species represents only those animals so closely related that they can reproduce among themselves. In our example, the species is represented as the sheet of paper containing the type of animal.

Some of the folders are thicker than others, since some contain many more species. For example, the folder labelled, "Insecta" must be enormous, since it holds about 900,000 species, while the folder labelled "Arachnida" only holds about 40,000 species. Research which file cabinet, drawer, folder, and divider would contain the sheet of paper for human beings.

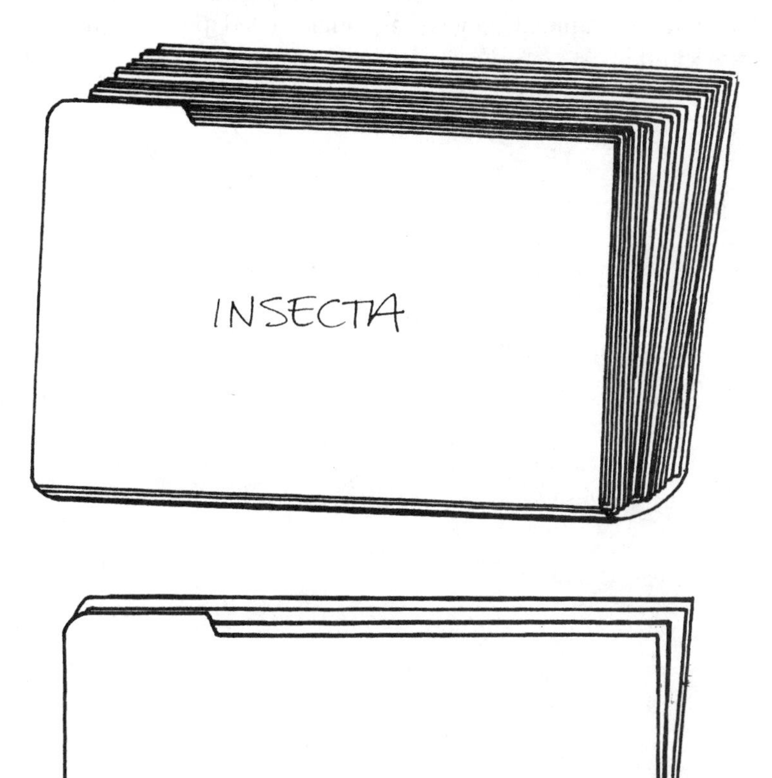

The animal kingdom is actually divided into over 20 phyla. Some of the more important phyla (and examples of each), including those represented in the experiments in this book are: Porifera (sponges), Cnidaria (jellyfish), Platyhelminthes (planaria), Nematoda (nematodes), Mollusca (slugs), Annelida (earthworms), Arthropoda (insects), and Chordata (animals with backbones, including people).

There are many different classification systems; some have three kingdoms, some four, some five. The most accepted system today uses five kingdoms. In this system, the animal kingdom only contains multicellular animals. Single-celled organisms, such as the paramecium and amoeba, are placed in the Protista kingdom, and bacteria are placed in the Monera kingdom. The projects in this book, however, include a wide range of "animals" including single-celled protists, such as the paramecium.

2

An introduction to scientific research

Science fairs give you the opportunity to not only learn about a topic but to participate in the discovery process. Although you probably won't discover something previously unknown to mankind (although you never can tell), you perform the same process by which discoveries are made.

Advances in science move forward slowly, with each experiment building upon a previous one and preparing researchers to perform the next. Advances in zoology, botany, microbiology, and virtually all scientific disciplines proceed one step at a time. A typical science fair project allows you to see what it is like to take one or two of these steps for yourself. The following typifies how science marches forward.

A problem, such as controlling a particular type of insect pest without harmful pesticides, might be solved by a series a scientific experiments. First, field studies could search for natural predators or parasites of the pest, or possibly natural products such as an extract from a local plant. Lab and field studies can then be performed to determine how effective each natural enemy or substance would be at controlling the pest. The life cycles of the predators and parasites or the side effects of the extract would be studied, as well as how each would fit into the local ecosystem.

Further studies might be performed to determine the population dynamics of the natural enemies or the proper concentrations of the extract. Are any of them capable of controlling the pest? What would happen to the entire ecosystem if the numbers of these predators or parasites were artificially increased or if a nearby pond became contaminated with the extract.

Some experiments might find that some of the enemies are incapable of controlling the pest or that the extract is poisonous to beneficial insects as well as the pest. Even these studies, however, are valuable since they provide information that keeps scientists on the correct track. Once a natural enemy or a plant extract is found to be a likely candidate, small scale testing can begin until a solution to the problem is found.

As you can see in the previous example, each experiment was necessary before the next could be performed, and the entire progression was necessary before a successful conclusion could be reached. Scientific research, no matter how simple or how sophisticated, must follow a protocol that demands consistency and, most importantly, duplicity. When one scientist or research team finds some new revelation, others must verify it. The scientific method provides a framework for researchers to follow. It assures a highly focused, reproducible sequence of events. The fundamentals of the scientific method are discussed next.

The scientific method

The scientific method can be divided into five basic steps. The following paragraphs describe each step and relates them to sections in this book.

Problem

What question do you want to answer, or what problem do you want to solve? For example, can a barrier of crushed egg shells prevent slugs from attacking garden plants, or does sight deprivation enhance a person's ability to hear? The project overview section of each project in this book gives a number of questions and problems to think about. The going further and suggested research sections in this book can also give you ideas to spark your imagination about additional problems to think about.

Hypothesis

The hypothesis is an educated guess, based on your literature search, that offers a possible answer to your questions. You might hypothesize that crushed eggs can prevent slugs from damaging garden plants or that sight deprivation does not actually enhance your ability to hear. You should form your hypothesis about these questions before proceeding with the project.

Experimentation

The experiment is designed to determine whether the hypothesis is correct or not. Even if the hypothesis wasn't correct, a well-designed experiment helps determine why it wasn't correct.

There are two major parts to the experiment. The first is designing and setting up the experiment. How must you prepare the experiment, and what procedures must be followed to test the hypothesis? What materials will be needed? What live organisms, if any, are needed? What step-by-step procedures must be followed during the experiment? What observations must be made and what data must be collected during the experiment? Once these questions have been answered, the actual experiment can begin.

The second part is running the experiment, making the observations, and collecting the data. The results must be documented for study and analysis. There are three important things to remember when performing research. Take notes, take notes, and take notes. The more details, the better. The most common mistake a new scientist makes is thinking he or she will remember some small detail. If you always carry a notebook (journal) and pencil when working on your project, this won't become a problem. Some science fairs require the project notebook be submitted along with a brief abstract of the project. Some fairs require or encourage a full length report of the project as well. (See the science fair project guidelines later in this section.)

The materials list section of each project lists all the materials needed for each experiment, and the procedures section gives step-by-step instructions. Suggestions are given on what observations should be made and what data should be collected.

Another aspect of experimentation is the importance of replication. For any project to be considered valid scientific work, the experimental groups should be replicated as many times as is practical. The duplicated groups can then be averaged together or, better yet, statistically analyzed.

For the projects in this book, try to perform all experimental groups in triplicate, even if not instructed to do so. For example, if you are collecting samples from a site, collect three times in the general area, or if you are running time trials of some sort, run them at least three times. Replication reduces the chances that incorrect conclusions will be drawn from the data.

Analysis

Once you have completed the experiment and collected the data, you must analyze it and draw conclusions to determine if your hypothesis was correct. Create tables, charts, or graphs to help analyze the data. The procedures section of each project suggests what observations to make and data to collect while running the experiment. The analysis section asks important questions to help you analyze the data and often contains empty tables or charts to fill in with your data. This book provides guidance, but you must draw your own conclusions. In the back of this book is an appendix that lists other books that provide information to help you analyze data.

The conclusions should be based upon your original hypothesis. Was it correct? Even if it was incorrect, what did you learn from the experiment? What new hypothesis can you create and test? Something is always learned while performing an experiment, even if it's how "not" to perform the next experiment.

Building on past science fair projects

Just as scientists advance the work of other scientists, so too can you advance the work of those that have performed other science fair projects before you. I don't mean copying their work, but thinking of what the next logical step might be to take in that line of research. Possibly you can put a new twist on a previous experiment. If electromagnetic radiation is harmful to single-celled organisms such as a paramecium, can it also harm a multi-cellular organism such as a nematode or a planaria? Or, if the original experiment was performed *in vitro* (in a test tube), can you perform a similar experiment *in vivo* (in nature)?

Abstracts of previous science fair projects are available from the Science Service in Washington D.C. See the book list in the back of this book for other sources of successful science fair projects.

A special note about human subjects

If you are using people such as your friends or family members as experimental subjects, be sure to explain fully to them what you plan to do, and get their permission before proceeding. If a student or child is participating, get written permission from their parent or guardian. It is unethical to include a person in an experiment, no matter how harmless, without their consent.

3

Getting started

The best way to select a project is find out what interests you about animals, if you don't already know. Did you ever wonder what hidden animals could be discovered in a handful of dirt or a cupful of pond water? Do you take vitamins? Can vitamins improve the regeneration capabilities of a starfish or planaria? Does someone in your family smoke cigarettes? Does second-hand smoke affect worms, insects, or people?

Stop and let your mind wander for a while. What comes to mind? It could be anything, anywhere, anybody! Once you've opened your mind and let your imagination run wild, look through the table of contents in this book for more specific topics to research. Select a project that you are not only interested in, but truly enthused about.

Since you are looking through this book, we can probably assume you have an interest in animals. Therefore, the first thing to do is find out specifically what peeks your interests, if you don't already know. There are a few ways to do this.

Use this book

The first thing to do is look through the table of contents in this book for topics to research. This book contains 20 science fair projects about animals. Read through the background and project overview sections of each project. Every project in this book can be adjusted, expanded upon, or "fine-tuned" in some way to personalize your investigation.

After reading through these sections, think about how you can put your own signature on the experiment. The going further and suggested research sections of this book are designed to help you do just that. If you find yourself saying, "I'd like to know more about . . ." something, you're well on your way to selecting a science fair project about animals.

Other sources

At this point, you can begin your project or continue to look for more insight into the problem. Consider branching out by looking through science sections of newspapers. Also, look at magazines such as *Discover* or *Omni*, which cover a broad range of topics. Check the *Reader's Guide to Periodical Literature*, which is an index that lists articles in numerous magazines and gives a brief synopsis of each. Your school textbooks might also be helpful. Check references to other books, usually found at the end of each chapter.

Other sources that can help include educational television shows such as *NOVA*, *Network Earth*, *National Geographic* specials, *Nature*, and many others. Almost all of these types of shows are found on public television or cable networks. Check your local listings to see what might be viewed in the near future in your area.

Talk to specialists in the field

Once you have a good idea for a project, consider talking with a professional. For example, if your project involves an insect pest, arrange to meet a professor of entomology at a nearby university and a farmer or a gardener who must contend with the pest. Also, try to meet with a chemist who understands the effects of the chemical pesticide used to control the pest. If the project is about the effects of second-hand smoke, speak with a smoker, a medical doctor specializing in respiratory ailments, and a research scientist studying the topic. Also, consider speaking with a representative from a tobacco company. Obviously you'll get many points of view. Then you can formulate your own conclusions. Interesting science fair projects don't only involve equipment, chemicals, and cultures, but also what people have to say about the topic: pro, con, and indifferent.

Also be sure to use any resources that are readily available to you. If you live near a zoo, aquarium, nature preserve, landfill, agricultural research station, large mechanized farm, small organic farm, or almost any type of facility that can contribute to your project, try to use it to your advantage. If you have a parent or friend who is involved in a business or profession applicable to your project, try to incorporate it into your research.

Put your signature on the project

All the projects included in this book have been successful science fair projects. Most went on to state ISEF finals. Although you could simply duplicate these projects, I suggest speaking with your teacher or sponsor about how you could modify these projects. What could make these projects outstanding examples of research is how you put your signature on them. You can include portions of the going further section or delve into the suggested research section. Did a teacher, scientist, or business person add an interesting aspect of the research to make it truly unique and your own?

Before you begin

Review the entire project with your sponsor to anticipate problems that might arise. Some projects must be done at a certain time of year. Some can be done in a day or two, while others can take a few weeks, months, or even longer.

Some projects use supplies that can found around your home, but many require equipment or supplies that must be purchased from a local hardware store, science/nature store, or a scientific supply house. Some projects require organisms such as planaria, mealworms, or lizards. Your sponsor might have access to the organisms needed for the project. Insects, such as mealworms, can be bought at a pet or bait shop or ordered from a scientific supply house. Colonies of nematodes or cultures of planaria might be available from your school, or they too can be ordered from a supply house.

Also, plan ahead financially. Look through the materials list section of each experiment. Be sure to add materials you need for additions or modifications you made to the original project. Determine how and where you will get everything, and how much it will cost. If a dissecting microscope is needed, do you have access to one? If a live animal is needed, can you catch it in your location during that time of year or must it be ordered from a supply house? How much will it cost? Don't begin a project unless you can budget the appropriate amount of time and money as suggested by your sponsor.

Performing the experiment

Once you have selected a project by following the suggestions in the previous chapter in this book, use the following suggestions to get organized.

Scheduling

Before proceeding, it is a good idea to develop a schedule to assure you have a complete project in time for the fair. Have your sponsor approve your timetable. Leave yourself time to acquire the equipment, supplies, and organisms.

Most science fair projects require at least a few months from start to finish if they are to be accomplished thoroughly. In many instances they can (and must) be completed in less time. It would be difficult to produce a prize-winning project, however, without plenty of time. Here is a general list of things to do for your project. Think about this when preparing a timetable.

- Identify your adult sponsor.
- Choose a general topic.
- Establish a project notebook (journal) for all note taking throughout the project.
- List resources (including libraries to go to, people to speak with, and businesses, organizations, or agencies to contact).
- Select reading materials and use bibliographies for more resources; begin a formal literature search.
- Select the exact project and develop a hypothesis.
- Write a detailed research plan and discuss it with your adult sponsor.
- Have your sponsor sign off on your final research plan.
- Procure equipment, supplies, organisms, and all other materials.
- Follow up on your resource list: speak with experts, make all contacts, etc.
- Set up and begin experimentation.
- Collect data and rigorously take notes.
- Begin to plan for your exhibit display.
- Begin writing report.
- Begin to analyze data and draw conclusions.
- Complete report and have sponsor review.
- Design exhibit display.
- Write final report and abstract, and be sure notebook is available and readable.
- Complete and construct dry run of exhibit display.
- Prepare for questions about your project.
- Disassemble and pack project for transport to fair.
- When the fair arrives, set up your display and have fun.

Literature search

As you can see from the suggested schedule, one of the first items is to perform a literature search of the problem you intend to study. A literature search (also simply called "research") means reading everything you can get your hands on about the topic. Read newspapers, magazines, books, abstracts, and anything related to the specific topic to be studied.

Use online computer services such as OnLine America, CompuServe, Prodigy, or Delphi. These offer an incredible opportunity to learn about a topic. They offer access to library bibliographies, online databases, and reference books. There is no better way to get so much information so quickly. Many of these services let you make requests for information using E-mail. The responses you receive might save you countless hours of searching for material.

Talk to as many people as possible that have some insight into the topic. Listen to the news on radio and television. At this point you might want to narrow down or even change the exact problem you want to study.

Once your literature search is complete and you have organized the data both on paper and in your mind, you should know exactly what problem you intend to study and then formulate your hypothesis.

The research plan

At this point you should have completed a final Research Plan. You can use portions of this book to get started with your Research Plan, but you must go into additional detail and include all modifications. Before beginning the project, go through it in detail with your Adult Sponsor to be sure the requirements of the project are safe, attainable, suitable, and practical. In many science fairs, your sponsor is required to sign off on the Research Plan attesting to the fact that it has been reviewed and approved.

It is important to review your fair regulations and guidelines to be sure your project won't run into any problems at the last minute.

Science fair guidelines

Almost all science fairs have formal guidelines or rules. Check with your sponsor to see what they are. For example, there might be a limit to the amount of money that can be spent on a project or the use of live (vertebrate) animals. Be sure to review these guidelines and check that the experiment poses no conflicts.

Many science fairs require four basic components for all entries.

- The actual notebook used throughout the project that contains data collection notes. Be sure to consider this when taking your notes since they might be read by fair officials.
- An abstract of the project that briefly states the problem, proposed hypothesis, generalized procedures, data collection methods, and conclusions. This is usually no more than 250 words long.
- A full length research paper (might be required).
- The exhibition display.

The research paper

A research paper might be required at your fair, but consider doing one even if it isn't necessary. You might be able to receive extra credit for the paper for one of your science classes. The research paper should include seven sections:

- Title page
- Table of contents
- Introduction
- A thorough procedures sections explaining what you did
- A comprehensive discussion section explaining what went through your mind while performing the research and experimentation
- A separate conclusion that summarizes your results
- A reference and credit section in which you list your sources and give credit to anyone who helped you (including any company, organization, or agency that assisted you).

Some books that detail how to write a research report are listed in the back of this book.

The display

The exhibit display should be as informative as possible. Keep in mind that most people, including the judges, will spend only a short amount of time looking at each presentation. Try to create a display that gets as much information across as quickly as possible with the least amount of words. Use graphs, charts, or tables to illustrate data. As the old saying goes, "A picture is worth a thousand words." Make the display as attractive as possible since you cannot communicate the value of your project if you don't draw peoples' attention to it.

Discuss with your sponsor any exhibit requirements such as special equipment, electrical outlets, and wiring needs. Live organisms of any kind are usually prohibited from being displayed. Often, preserved specimens are also prohibited. Usually no foods, wastes, or even water is allowed in an exhibit. Also no flames, gases, or harmful chemicals are allowed. Find out what you can and cannot do before proceeding.

Many fairs have specific size requirements for the actual display and its backboard. For more information on building an exhibit display, see the book list in the back of this book.

Judging

When beginning your project, keep these things in mind. Adherence to the scientific method and attention to detail are crucial to the success of any project. Judges usually want to see a well thought out project and a knowledgeable individual who understands all aspects of his or her project.

Most science fairs assign a point value to various aspects of a project. For example, the research paper might be worth 15 points while the actual display might be worth five points. Request any information that might give you insight about the judgment criteria at your fair. This can help you allocate your time and resources where they are needed the most.

Part 2

Studying the human animal

The study of zoology includes animals as simple as microbes and as complex as man. This first part contains four projects to help us learn more about ourselves.

The first project investigates whether video games can affect our cardiovascular activity and continues on to see if certain people are more prone to be affected by these games.

As we age, many of us begin to lose our hearing, but do we also begin to lose our sense of smell? This question is investigated in the second project.

Next, some people think that the blind have an uncanny ability to hear. The third project tries to determine if a person's ability to hear would improve if they were blindfolded for a short period of time.

The last project looks into the emotional and physiological effects of various types of music, from classical to rap, on human beings.

Video games & your heart

Is there a relationship between video games & heart stress?

Some people are more likely to become excited or angry over events that don't even seem to affect others, a phenomenon called coronary prone behavior. Extensive research has been performed over the past 30 years to study this phenomenon, and researchers have tried to determine the underlying cause and effect of this behavior. Early studies called this type A behavior and, more recently, high hostility behavior. Some of these studies indicate that people showing signs of coronary prone behavior are more prone to cardiovascular disease.

Project overview

This project investigates whether video games, especially those with a violent content, can lead to increased cardiovascular activity. It continues on to see if there is a link between individuals who are affected by the video games and individuals who score high on a coronary prone behavior test. If there is a link between these two groups, it would lend credibility to the hypothesis that people with coronary prone behavior have an increased risk of cardiovascular disease.

Materials list

- Nintendo Entertainment System and two video games (Any video game system with two video games—one containing non-violent action and one containing intense action—will work; the two video games used for the original project were Dr. Mario and Renegade.)
- Marshall F-89 Astrofinger hand-held blood pressure monitor or similar blood pressure monitor (You might borrow such a device from your school nurse or find one in a medical supply store.)
- 75 subjects willing to participate
- Two psychological tests that evaluate coronary prone behavior (The original project used the Cook and Medley Hostility Scale and a self-test for type A personality from Stress/Unstress, 1981, by K. W. Schnert, M.D. Other tests are available; check with your sponsor.)

Procedures

You need at least 75 subjects to participate in this study to make it statistically significant. All the subjects should fall within a certain age range, for example ages 12 to 17. Although this test is perfectly safe, you should prepare a written consent form for the students and their legal guardians to sign.

A testing area should be set up in an enclosed room with the video game system and blood pressure device present. The physiological testing will be done in this room. Each person participating in the project will complete three phases of physiological testing. When they finish these three phases of physiological testing they will take psychological tests.

In phase one of the physiological testing, subjects rest for two minutes and then engage in play on an intense video game. (See Fig. 4-1.) After three minutes, the subject stops playing and physiological data is collected. The physiological data is collected using the Marshall F-89 Astrofinger hand-held blood pressure monitor or a similar device. The subject's systolic blood pressure, diastolic blood pressure, and heart rate are recorded according to the instructions provided with the device.

Phase two is identical to phase one, except a non-violent game is played by the subject. Following phase two, the same physiological tests are administered as previously described. In phase three, the in-

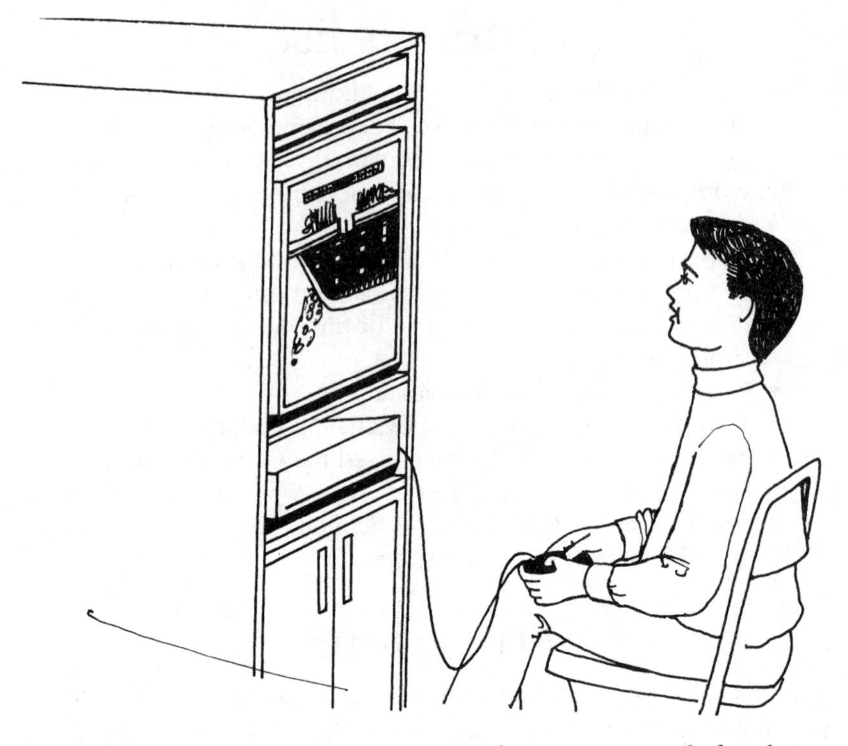

4-1 *The subject observes a video game for two minutes before being tested.*

dividuals simply remain at rest, playing no video games, while the physiological data is collected. The individual should be in a quiet setting with no distractions for five minutes before the blood pressure and heart rate measurements are taken.

After all 75 individuals have finished all three phases, the psychological testing begins. Each subject completes two psychological tests to evaluate for coronary prone behavior. There are many such tests available. You can use any appropriate test, but the original project used a self-test for type A personality, reprinted with permission from K.W. Schnert, M.D., from his book *Stress/Unstress*, (1981) and the Cook and Medley Hostility Scale, by Redford Williams, M.D., which is a 50 question test. (If you are going to photocopy these tests, you must obtain written permission from the publishers.)

Analysis

To compare and analyze the physiological data, you should calculate the mean blood pressure for each individual for each of the three phases. Mean blood pressure is the average pressure throughout the cardiac cycle, and it can be approximated using the following formula:

$$\frac{\text{SYSTOLIC-DIASTOLIC}}{3} + \text{DIASTOLIC READING} = \text{MEAN PRESSURE}$$

Did some people display a greater cardiovascular reaction to the video games than others? Can you categorize groups of people who were more prone to show a physiological change from playing the video games? Did these people respond differently to the two video gamess, or did they respond in the same way?

Next, compare the physiological results with the results of the psychological tests. Is there a correlation between those people who showed signs of coronary prone behavior and those who showed significant reactions in the physiological tests? If a link was found between these two tests, what can you conclude from this data? How might this information be used to help reduce the risk of coronary disease in these people?

Going further

Run a series of statistical analyses, such as the student's t-test and a paired t-test, to see if the differences among the various groups were significant. Continue the project to see if there are differences between males and females in both tests and in the final analysis. Survey the participants to find out if any have coronary disease in their immediate family. Since this disease is often hereditary, how does this tie into your results?

Suggested research

- Investigate the recent literature about the link between coronary disease and what is commonly called type A behavior.
- Look into the various techniques used to help people who are prone to coronary disease handle stress. What approach do they take?

5

Age & smell

Does aging affect your sense of smell?

As people age, they begin to lose some of the abilities they once had. Overall movement might be reduced, the ability to remember things might be compromised, and quick reactions are slowed. Our senses are also not as keen as they once were. Glasses are often needed to correct some loss of vision, and a hearing aid might be needed to boost sound. Do people also begin to lose their ability to smell things?

Project overview

During this project you will use at least ten different common odors. Each odor will be inhaled from a small canister made from a old film canister with mesh covering. The subjects will be asked questions, and their responses will be recorded on test forms. (See Fig. 5-1.) These forms indicate the individual's sex, age, and age category and the number of correct and incorrect responses that they made during the test.

The odors should include, but are not limited to, cinnamon, peppermint, ammonia, garlic, talcum powder, orange, coffee, bathroom cleanser, rose, and moth balls. Each age category should include ten people and consist of five males and five females. Since there are eight age categories, you'll need about 80 people for the test. (The older categories will probably have fewer people. You might want to group the two oldest categories, 71 to 80 and 81+, together to form one category.)

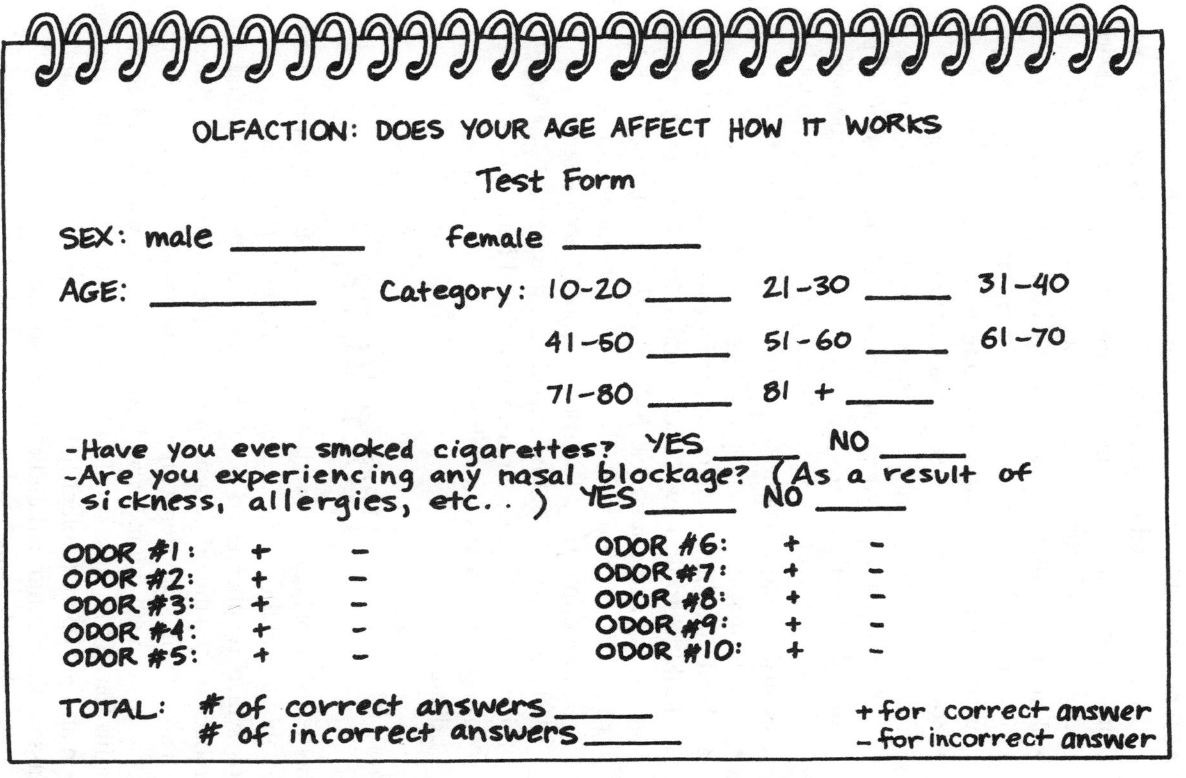

OLFACTION: DOES YOUR AGE AFFECT HOW IT WORKS

Test Form

SEX: male ______ female ______

AGE: ______ Category: 10-20 ____ 21-30 ____ 31-40

41-50 ____ 51-60 ____ 61-70

71-80 ____ 81 + ____

-Have you ever smoked cigarettes? YES ____ NO ____
-Are you experiencing any nasal blockage? (As a result of sickness, allergies, etc. .) YES ____ NO ____

ODOR #1: + −
ODOR #2: + −
ODOR #3: + −
ODOR #4: + −
ODOR #5: + −

ODOR #6: + −
ODOR #7: + −
ODOR #8: + −
ODOR #9: + −
ODOR #10: + −

TOTAL: # of correct answers ____
# of incorrect answers ____

+ for correct answer
− for incorrect answer

5-1 *You will fill out a form such as this one for each subject being tested.*

Materials list

- Ten film canisters (black or non-transparent)
- Ten film canister covers
- Ten 3-inch-diameter pieces of mesh to cover each canister (The mesh must be opaque to conceal the jar's contents, yet it must be porous enough to allow the odor to escape. Dark cloth or black pantyhose will work.)
- Ten rubber bands (to hold mesh securely on canisters)
- Twenty stickers/labels (Since the film canisters are black, you won't be able to write a label directly on the canister.)
- Sources for ten odors, such as cinnamon (use a powder), peppermint (crushed mint disks or candy cane), ammonia, garlic (or garlic powder), talcum powder, orange (orange peel and slices, with juice), coffee (fresh ground), bathroom cleanser (Comet, Ajax, etc.), rose (rose fragrance or actual petals), or naphthalene (moth balls)
- A scent-free room for the testing
- About 80 subjects willing to participate in the tests (You should try to find five males and five females in each of the following age categories: 10–20, 21–30, 31–40, 41–50, 51–60, 61–70, 71–80, and 81+. The last two categories can be combined into one. The subjects should not be smokers. Also, they should not be experiencing any nasal blockage due to sickness or allergies when tested and should not be using nasal sprays since they cause the olfactory bulb to dilate.)

Procedures

Label each "odor" canister with a number. Place a stick-on label onto the canister and then write the label on the sticker. Note in your notebook which number refers to each odor. Canisters are labeled with numbers to keep the odor name from being revealed to the participants.

Fill each film canister with its respective odor source. Then cover each canister with the mesh and hold it in place with a rubber band. Do this for all the canisters. The best time to run this test is between noon and 3:00 p.m. Replenish the "odor" canister contents each hour to assure their continued and constant strength.

Once the odors are ready, you can begin to test the participants. Fill out a test form for each person before administering the test. Then have the person smell canister #1 for five seconds. Take one and only one response for each odor, and note whether the response was correct on the form as shown in the figure. Tell the participants you will only accept answers that are specific. For example, do not accept "flower" for "rose."

Leave two minutes between each odor tested to clear out each participant's olfactory membrane. The remaining odor molecules from the previous odor must have time to leave the nasal cavity. (See Figs. 5-2 and 5-3.) Everyone should smell the odors in the same order.

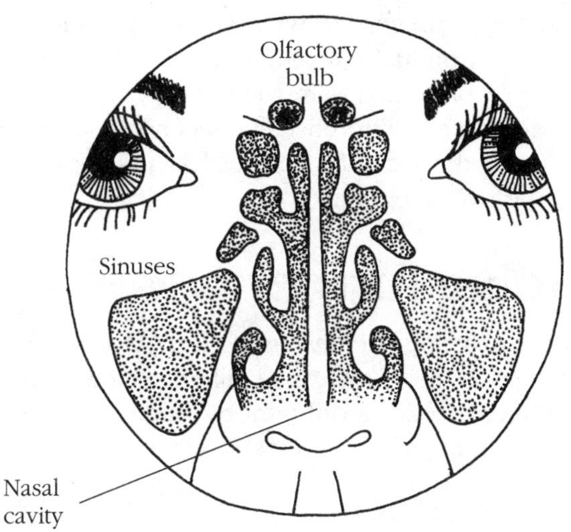

5-2 *The olfactory bulb plays an important role in a person's ability to smell.*

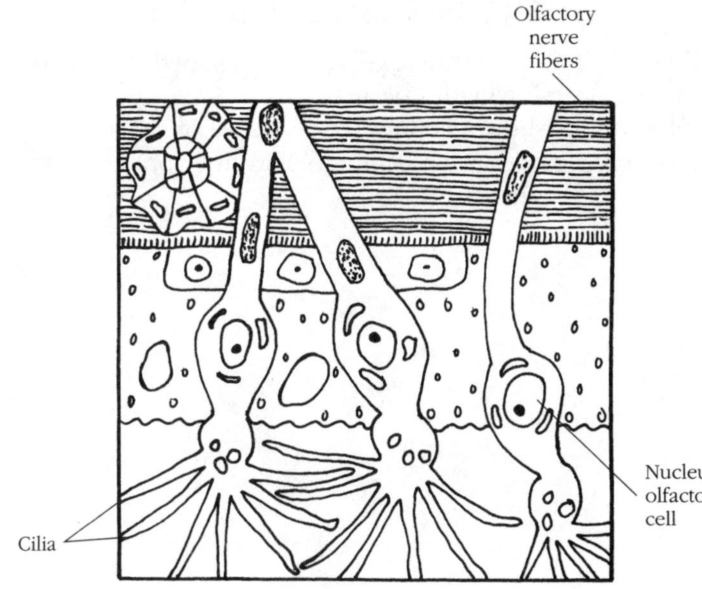

5-3 *Here is a closer look at the olfactory anatomy.*

You will probably have to continue testing on different days to test all the subjects. Be sure to keep all conditions the same throughout the experiment. Don't change rooms or any other factors.

Analysis

First, calculate your results for each category and then compare categories. Do the number of correct responses decline as the age category increases? Is there an age category in which there is a dramatic loss in smell or does it occur gradually throughout all the age categories? Plot your data in a graph to compare each group.

Study the data to compare the differences between the sexes in each category. Does it appear that one of the sexes has a better sense of smell in any or all of the age categories?

Going further

Continue the project by analyzing the results collected on certain odors. Are some types of odors more or less readily detected by different age groups or one gender? For a different twist, devise an experiment that will test to see if dogs begin to lose their sense of smell as they grow older.

Suggested research

- If you found one of the sexes to have a superior sense of smell, research why this occurs.
- If the results show a definite loss in sense of smell with age, investigate exactly what is physiologically happening that makes this occur.

6

Sight deprivation & hearing

Does short-term loss of sight enhance your ability to hear?

It is often said that blind people can hear better than people with sight. Some people are confident that physiological changes must occur in blind people, and those changes allow them to hear better than most people. Current research, however, indicates that this improved hearing is due to behavioral adaptation, not physiological changes. This adaptation allows blind people to focus their concentration on sound, and they therefore appear to have a more acute sense of hearing than sighted people.

If this is true, is it possible to observe an increased ability to hear in a person who is blindfolded? How long would it take before this person was able to hear better than before being blindfolded?

Project overview

The purpose of this experiment is to determine if short-term sight deprivation has any effect on the hearing of individuals whose hearing is classified as "clinically good." What would happen if a person with clinically good hearing was deprived of sight for short periods of time? Would their hearing improve?

In this project, you will evaluate the hearing of at least fifteen individuals with and without the stimuli of vision. You will first test subjects with their eyes open to determine the lowest decibel level at which they can hear. Then the subjects will be blindfolded and the same test will be performed immediately after blindfolding and again at intervals of five, ten, and fifteen minutes. Finally, you will measure the differences between the groups and note the patterns of hearing improvement and regression.

Materials list

- Beltone model 112 audiometer (Any audiometer will work. You might be able to borrow one from a local audiologist. Be sure to get instructions about how to use the device.)
- Blindfold
- Clock or stopwatch
- At least 15 subjects including all ages and both sexes (All subjects must be able to hear at 20 dB or less, which is considered "clinically good" hearing)
- A silent testing location

Procedures

Hearing tests must be run under the same conditions. Be sure the room is completely void of outside noises and other kinds of distractions.

First, you must test the subjects to see if they have clinically good hearing. To do this, have each subject keep his or her eyes open, and use the audiometer to test hearing at 20 dB. All tests should be run at 1000 Hz. (See Figs. 6-1 and 6-2.) Most audiometers have an indicator button, which the subject presses to indicate whether a sound was heard.

The audiometer used for the original project was set to "tonal input" and masking set to "off." Discuss the settings to be used on your device with the owner of the device before proceeding with the project. All tests are conducted on only one ear. (The original project used the left ear only.) Since this experiment is concerned only with changes in hearing, the ear used has no effect on the results.

Any subject who does not indicate that he or she has heard the tone at 20 dB must be rejected and replaced with another. For those subjects that hear 20 dB, record their sex and age and continue the experiment.

If the subject has clinically good hearing, continue with the subject's eyes open. Sound a tone between 40 and 70 dB to give the subject an idea of what he or she is listening for. Then establish the

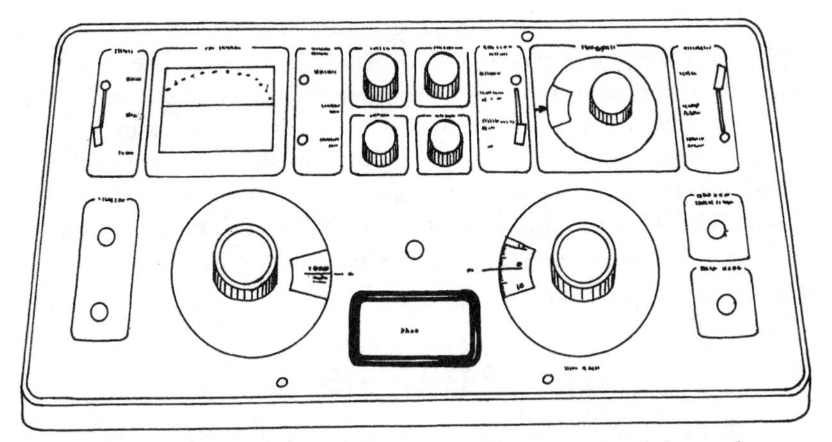

6-1 *You will use an audiometer to record a person's ability to hear.*

6-2 *Most audiometers come with a headset and a signaling button.*

subject's base level of hearing with sight. In other words, with the subject's eyes open, test for the lowest intensity he or she can hear. Begin at –10 dB, and increase the intensity of the tone by one decibel at a time until the subject acknowledges having heard the tone. Record the lowest intensity (in decibels) that the subject was able to hear with sight.

You can now move on to the test phase of the project. Blindfold the subject and repeat the same test immediately, as stated above. This time, you will try to establish the lowest intensity heard when the subject has just been deprived of his or her sight. Record the results. Wait five minutes and repeat the test again (the blindfold is still on the subject).

Repeat this step two more times at 10 and 15 minutes after being blindfolded. After the final test (at 15 minutes) the subject can remove the blindfold, and the testing is complete for this subject. Repeat the test with all of the subjects.

Analysis

After all the subjects have been tested, calculate the difference in decibels between the first and last tones recorded for each person while blindfolded. Did hearing improve when people were blindfolded? If so, how much of an improvement occurred? (An improvement of 10 decibels is equivalent to a 100% improvement.) Plot graphs for each subject tested. (See Fig. 6-3.) Then create a summary plot for all the subjects.

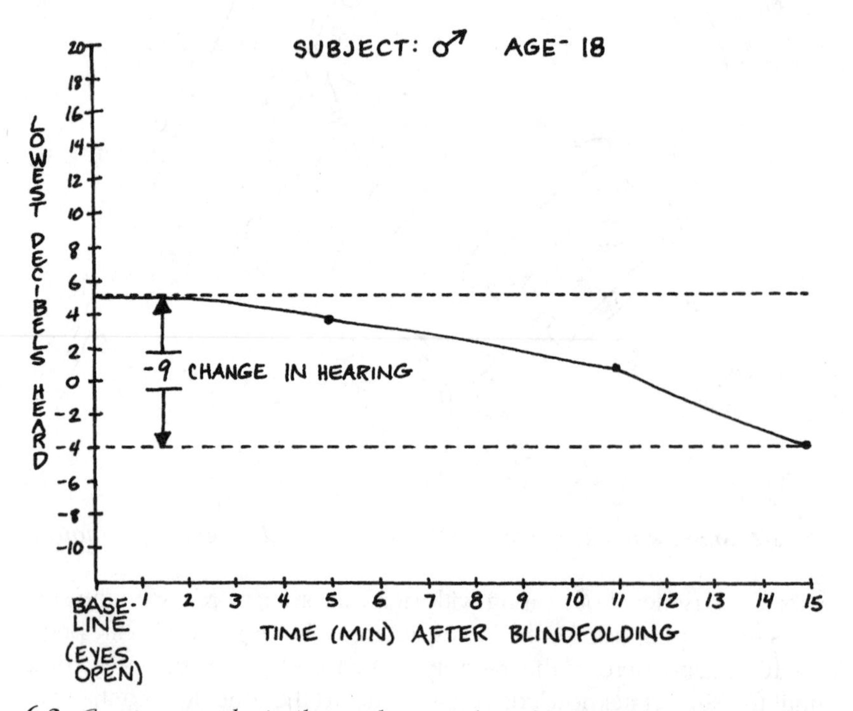

6-3 *Create a graph similar to this one that plots the change in hearing after being blindfolded.*

If hearing did improve, was it gradual or sudden? Did it change in all groups of people (gender and age) or only in some groups? Obviously, no physiological differences could have occurred in 15 minutes, so what might have caused any changes you observed?

Going further

What would happen if the subjects were distracted in some way while taking the blindfolded tests? Would the results be the same if the subjects were touched with a feather while being tested? Another way to continue the experiment would be to concentrate on different groups of subjects. Do you find significant differences between sex or age?

Suggested research

- Read the most recent literature about sight deprivation and hearing.
- Look into a relation between this topic and savants.

7

Music & the body

Do different types of music affect our bodies differently?

Many illnesses are believed to be brought on by stress. Some people believe you can prevent or even cure illnesses, especially those caused by stress, with thoughts, emotions, and a positive overall attitude. Different methods are used by different people to achieve a certain state of mind to encourage a healthy body. Music is often involved in this process. Some people chant, while others sing. Music has long been considered the soother of the soul.

If the mind and body are so closely linked to one another, it should be possible to see how one affects the other. This project does just that. It tests to see if different types of music can affect the body differently.

Project overview

The purpose of this experiment is to test the effects of music on the mind and body. What impact does music have on a person's vital signs? Do different types of music elicit different responses?

This project determines what effects, if any, different kinds of music have on a person's physiology. Can music affect a person's pulse rate, blood pressure, galvanic skin response, and overall emotions? Do different musical styles create different responses? Will everyone react in the same way to certain types of music, or will the responses be unique for each person tested?

Materials list

- Radio
- Stethoscope (probably available from your school nurse)
- Sphygmomanometer, which is a blood pressure cuff (probably available from your school nurse)
- Galvanometer, which measures the potential amount of electricity that can pass through your skin (possibly available in your science department)
- Three electrical wires, about eight to ten inches long
- Two alligator clips
- One 6-volt battery
- Variety of music (about 12 diverse songs)
- At least 25 participants and preferably more
- A room to perform the tests
- Writing pads for each participant

Procedures

Before beginning, you must get baseline data for each participant. To take "resting" (no music) measurements for all the participants, have each participant sit quietly in a quiet room for three minutes. Then take his or her blood pressure using the sphygmomanometer. (See Fig. 7-1.) Have the nurse show you how to use this device and read the results. Next, take his or her pulse. Again, have the nurse show you the proper technique.

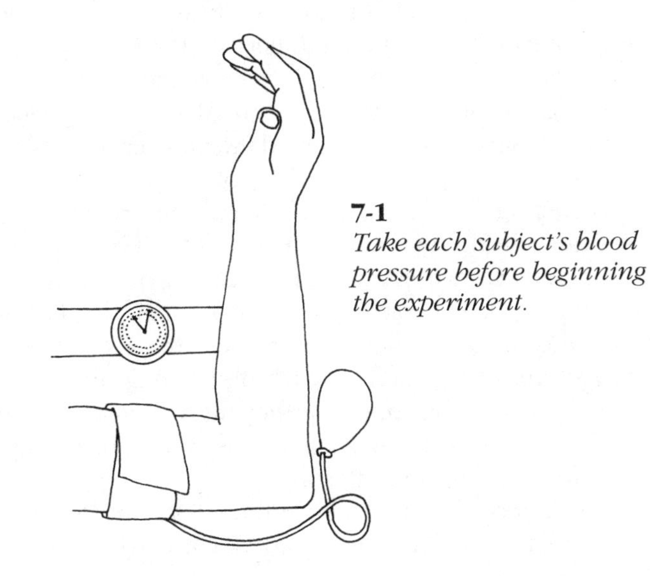

7-1
Take each subject's blood pressure before beginning the experiment.

The next test is the resting galvanic skin response. To perform this test, you must first assemble the galvanometer. Begin by preparing the wires. Attach one alligator clip to one end of one of the wires. Attach the other end of this wire to one of the terminals on the galvanometer. (See Fig. 7-2.) Attach the other alligator clip to one end of the second wire. Attach the other end of this wire to a terminal on the battery.

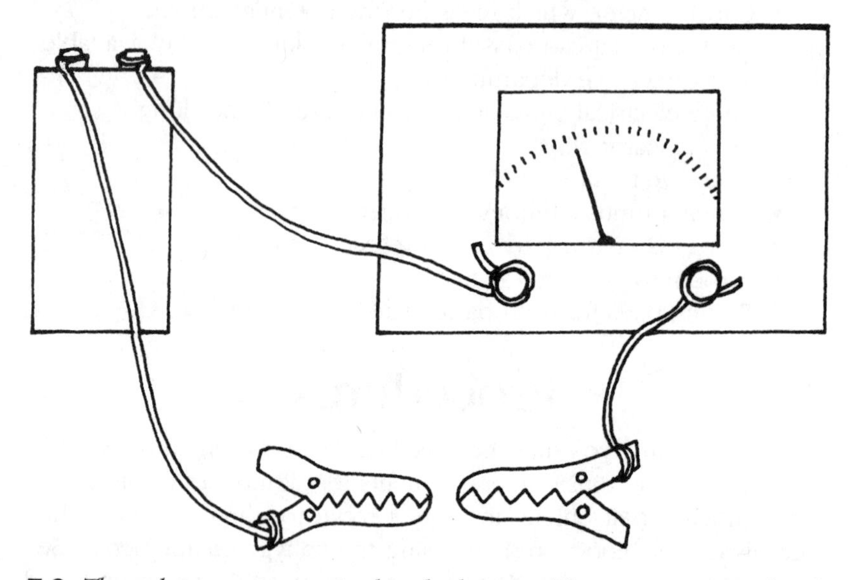

7-2 *The galvanometer is wired to the battery. The circuit is completed when a person holds both alligator clips.*

Take the third wire, which has nothing on it, and attach one end to the remaining battery terminal and the other end to the remaining galvanometer terminal. Now that the device is ready, have the first participant hold the alligator clips, one in each hand. All the participants must hold these clips in a similar way. Then read the number of units on the galvanometer. The results are measured in milliamps. Record the results. Before preceding, have your sponsor check this circuit for safety.

Now you can start the experimental part of the project. Play the first selection of music. Two minutes into the song run all three tests once again. Also, while the participants are listening to the music, have them write down how they feel on the writing pad. The pad should have a section designated for each of the songs to be played. Ask them to write down what emotions they feel at that time. The paper you provide can have a list of emotions from which they can check off their feelings during each song.

After you have recorded data for the three tests (and participants have filled out the form for the first song), begin playing the second song. Once again wait two minutes and begin the next set of tests and have the individual fill out the form for the next song. Repeat these steps for at least five diverse songs including everything from classical to rap music. Repeat the entire procedure for each participant in the test.

Analysis

After all the data has been collected, plot a bar graph for each participant that shows the responses for each song tested (see Fig. 7-3). Then compare the data for all the participants for all four physiological tests.

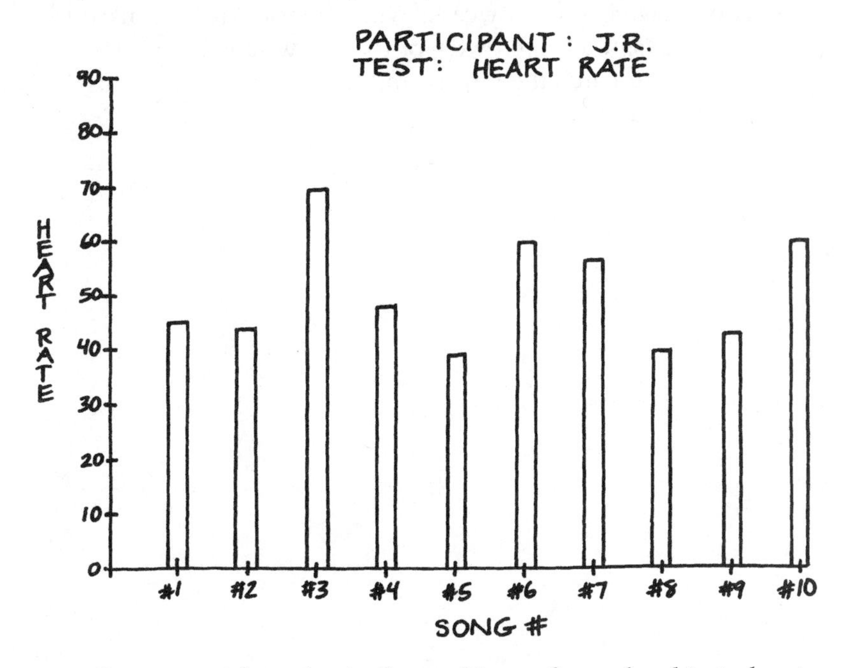

7-3 *Plot a series of graphs similar to this one for each subject, showing changes in heart rate and other factors for each song.*

Do some people appear to be affected by music? Do some types of music elicit greater responses than others? After studying and analyzing the data for each participant, see if similarities can be found among the participants. Can you find any types of music that affect everyone tested?

Going further

If music did affect at least some people, continue the test to see if any specific groups are more effected than others. For example, do females or males show more of a response to a certain kind of music? Or, are certain groups less likely to show a response to any music or a certain type of music? Did any one physiological test show the greatest change? What might this mean?

Suggested research

- Research psychosomatic illnesses. What does current medical literature say about the relation between the mind and the physical state of the body?
- There have been many cases where music has been used to incite riots, to calm crowds, or to otherwise influence the masses. Research the history and science involved in this phenomenon.

Part 3

Biocides

Most people call them pesticides, but the term is a misnomer: the term "pesticides" leads people to believe that they only kill pests. The vast majority of pesticides don't know the difference between an organism that is a pest and one that is not. For this reason, some people believe the term "biocides" should replace pesticide. Biocide means to kill life.

The world is awash in biocides. Traces of biocides have been found in the polar ice caps and can be found in mother's breast milk. Rachel Carson's famous book *Silent Spring* was the first to forewarn of impending problems with these chemicals. Today, the facts are clear. Ecosystems are threatened and so too are peoples' health.

With this new concern has come a deluge of research to find safer alternatives to dangerous pesticides. How can we protect ourselves and our crops from pests without polluting our environment and ourselves? This section includes five projects that look for answers to this dilemma.

The first project in this part investigates whether extracts from plants can be used as natural pesticides. The second project looks into ways to control a common garden pest, the slug, by attracting and trapping them. The third project examines one of the most promising alternatives to pest control—biological control, which is the use of predatory insects to control insect pests. The fourth project here investigates whether common herbs have any anti-microbial qualities that could be used to control microbes that spread disease.

The final project in this part returns to the common garden slug. This time, however, the research examines how to prevent the slug from getting into your garden. Can physical barriers, such as crushed egg shells or wood chips, keep slugs from eating your garden vegetables?

8

Natural pesticides

Do plant extracts contain any natural pesticides?

The vast majority of pesticides are synthetically produced, meaning we make them up in the lab. Since they are artificially produced, either they don't break down naturally or, when they do, they often break down into substances just as harmful as the original compound. Natural substances can also be dangerous, but almost all of them readily break down into harmless substances to be recycled through the biosphere once again.

Recently, there has been a new push to find *organic* (naturally occurring) substances to control pests, because these substances are usually far less harmful to the ecosystem.

Project overview

The purpose of this experiment is to see if there are any naturally occurring insecticides (a type of pesticide that kills insects) in plants such as cumin or garlic. The common fruit fly pest, Drosophila melanogaster, is used as the test subject.

There are two parts to this project. In the first, you will determine whether these plants have any pesticidal qualities. To do this, you'll see if any of the extracts can kill the fruit flies in an exposure test. In the second part of the project, you'll see if these extracts can act as a repellant or an attractant to the flies in an avoidance test.

Materials list

- Five grams of each of the following plant products (you can add your own to this list): cumin seeds, fennel seeds, ground black pepper, pepper corns, crushed red pepper, dried garlic (not powder), fresh garlic
- Mortar and pestle to create the extracts
- Test tube for each extract
- Parafilm
- Screw cap vial (about 30 ml) for each extract (for Set One as described in the project)
- Acetone (available at hardware stores)
- Fume ventilation hood
- Corn oil
- Calibrated pipettes to measure out the extracts
- Beaker (200 ml) to mix the acetone and oil
- Screw cap vial (about 1.5 ml) for each extract (for Set Two as described in the project)
- Glass bottle with screw caps (about 350 ml) for each extract used (for Set Three as described in the project)
- Graduated cylinders (10 ml and 25 ml)
- A colony of *Drosophila melanogaster* (These are available from scientific supply houses; you'll need adults for the exposure test and the third instar larvae for the avoidance test.)
- Refrigerator to cool the flies
- Screw cap plastic vial to hold recovering flies
- Cotton balls
- Instant *Drosophila* growth medium
- Gelatin
- 9-mm plastic petri dishes containing agar for each extract (agar plates can be made from a mix or purchased prepackaged from a scientific supply house)
- Filter paper
- Paper hole puncher

Procedures

Part one—the exposure test

First, you must prepare an extract of each plant material to be tested. To do this place 5 g of each plant material into a mortar bowl and add 25 ml of acetone. Use the pestle to grind the material thoroughly. (See

Fig. 8-1.) After the grinding is complete, let the acetone and plant material soak for 24 hours. Most of the acetone will evaporate, so place the mixtures under a fume ventilation hood during this period. After 24 hours, pour the mixture into a test tube and cover with parafilm. Then, pour each mixture from the test tube into a vial. Use the screw cap to create a tight seal so the remaining solution does not evaporate. (This is Set One, as described in the materials list.) You will have one sealed vial for each extract. Be sure to label each of the vials.

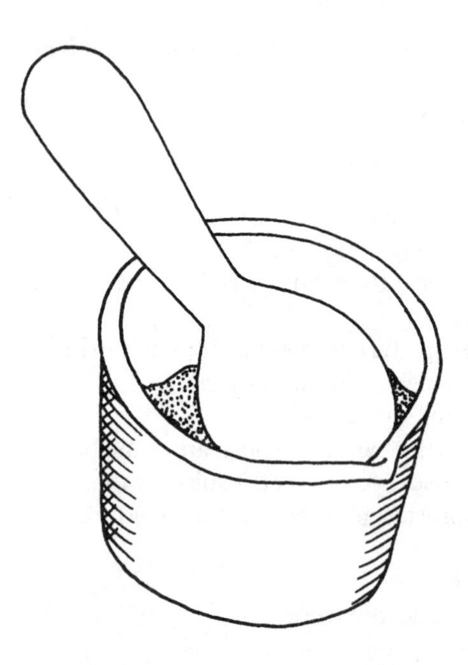

8-1
Use a mortar and pestle, available at your school lab, to create each extract.

You will now mix the plant extracts with an acetone and corn oil solution. To prepare this solution, mix 100 ml of acetone with 1 ml of corn oil in a beaker. Label a set of new vials with the name of each extract, plus label one "control." (This is Set Two, as described in the materials list.) Place 0.5 ml of this acetone/corn oil solution into each of these vials. Then, add 1 ml of each plant extract to each of the vials containing the acetone/corn oil solution. In the control jar, add 1 ml of acetone in place of extract.

Now that the extract solutions are ready for use, you'll prepare the test bottles. Add each extract mixture (extract/acetone-oil solution) to each 350 ml bottle and rotate it slowly to coat the surfaces evenly. Don't cover the bottle. Allow the acetone to evaporate for 15 minutes. This will leave the oil with the extract coating the bottle.

To keep the flies (which are about to be placed in the bottles) from becoming dehydrated during the tests, add a two cubic cm cube of gelatin to each test bottle. You should now have one test bottle containing a gelatin cube ready for each extract.

You will now place flies into each bottle to see if any of the extracts are capable of killing the flies. The procedure for exposing the flies to the test bottles is as follows. Anesthetize the flies by chilling them in a refrigerator for approximately ten minutes. Count out the anesthetized flies into groups of twenty. Place each group of 20 flies into a culture vial to recover. As they begin to recover, place the 20 flies into each extract bottle. As soon as the flies have been placed into their respective exposure bottles, begin timing the experiment. Record how many flies are found dead every hour for approximately 24 hours. This will tell you whether the substance being tested is toxic and will give you an estimate of its degree of toxicity to the flies. Repeat this procedure for each extract.

Part two—the avoidance test

Once you know which of the extracts appears to have some pesticidal qualities against the flies, you'll test to see if the flies will try to avoid the extracts. Even if an extract does not kill a pest, it can still be an effective repellant. This is how many insect sprays work. They don't kill the pest, they just keep them at bay. Alternatively, if the pest is attracted to the toxic substance, it is more likely to have a chance to kill the pest.

The procedure for larval avoidance is as follows. Cut the filter paper to fit (if it doesn't already fit) into the 9-mm-diameter petri dishes. Use a paper hole puncher to punch out two small holes in the filter paper and label. (See Fig. 8-2.) Place the paper in the petri dish on the agar and apply one drop of an extract into the appropriate hole. (The agar will keep the liquid extract in the hole.) Wait for a few minutes to allow the acetone to evaporate. Add a drop of distilled water to the other hole on the opposite side of the filter paper.

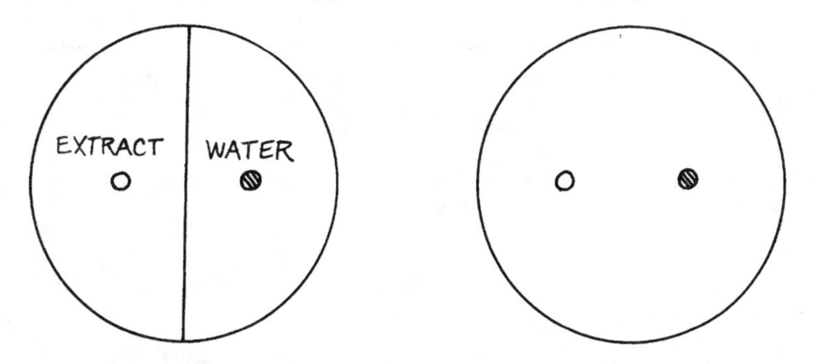

8-2 *Create one small hole for the extract and one for water during the larval avoidance test.*

To begin the avoidance test, add 20 third instar *Drosophila* larvae to the center of the petri dish directly between the two holes. (Third instar larvae are the largest larvae still found in the growth media.) As soon as the larvae are placed in the petri dish, begin timing the experiment. After five minutes, record the number of larvae on each side of the filter paper. Repeat this procedure for each of the extracts. Are the flies clustering around the extract or the water or are they randomly distributed around each?

Analysis

In the first part of the experiment, did any of the extracts kill the fruit flies? If so, how long did it take? Do any appear to have the potential to be used as insecticides? Plot a graph to show the mortality rate over time for each extract. (See Fig. 8-3.) In the second part of the experiment, did the larvae tend to avoid any of the extracts, or were they attracted to them?

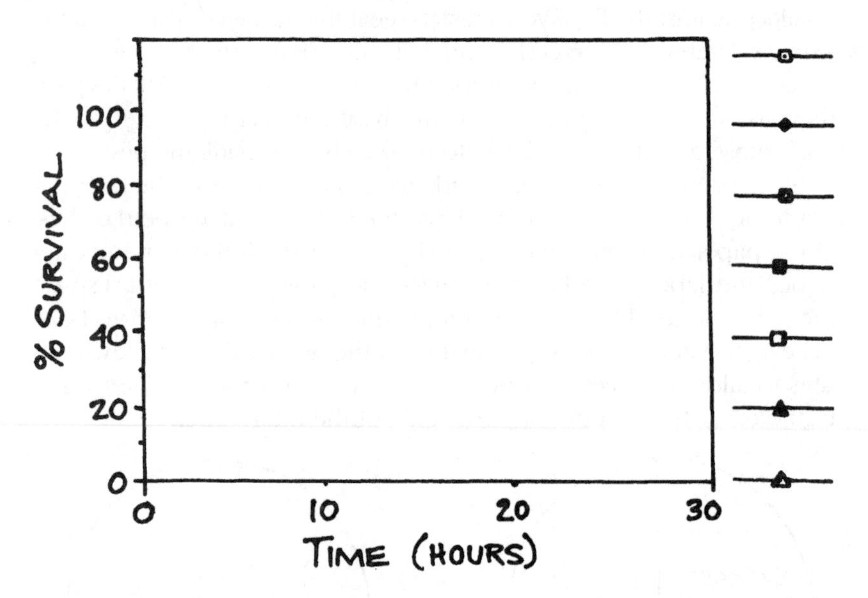

8-3 *Plot a graph similar to this one showing the death rate of the subjects over time for all the extracts.*

You can establish an "Avoidance Index" for each extract with the following formula:

Avoidance Index = [(# of flies around control drop) – (# of flies around extract drop)]/Total # of flies.

Were those extracts with the highest avoidance index the same ones that killed the flies in the exposure test?

Going further

For those extracts that killed the flies in the exposure test, determine their degrees of toxicity. Do this by diluting the extracts and repeating the first experiment. Will one-half or one-quarter concentration still kill the flies? The procedure for diluting the extracts is as follows. To create a one-half strength extract, add 0.5 ml of acetone to 0.5 ml of extract. Then add 0.5 ml of the .5% corn oil/acetone solution. Use this to coat a test bottle marked "½ strength."

To create a one-quarter strength extract, add 0.75 ml of acetone to 0.25 ml of extract. Then add 0.5 ml of the corn oil/acetone solution, and use to coat test bottles labelled "¼ strength." Add flies to these new test bottles and count the number dead flies over time. How do they compare with the full strength tests?

During the avoidance part of this project, you tested the third instar, larval stage. Continue the project to determine avoidance to the extracts by the adult flies.

Suggested research

- Look into organic gardening literature that advocates the use of natural pesticides. What types of botanical (plant) pesticides do they recommend?
- Contact organizations such as the National Coalition Against the Misuse of Pesticides in Washington, D.C. to receive literature about the problems associated with synthetic pesticides.
- Contact chemical companies that manufacturer pesticides to get their point of view.

9

Slug prevention

Can a mechanical barrier keep slugs out of your garden?

There are many alternatives to using synthetic chemical pesticides to kill pests. Although often overlooked and underutilized, mechanical means of pest control can be effective. The use of traps, barriers, lights, sounds, and other physical devices are becoming more popular as people realize the harm caused by chemical pesticides.

Slugs are common pests of gardens and greenhouses. They eat leafy vegetation and can cause considerable harm if left unchecked. Since they are slow moving creatures and easy to catch, a simple form of mechanical control is to use a forceps to pick the slugs off the plants.

Project overview

An even better method of protecting your garden plants from slug damage is to prevent them from ever getting on the plants in the first place. Because slugs cannot fly, the only way they can move into a garden is by crawling. An inexpensive, practical, and environmentally safe way to prevent these pests from infesting your garden would be to use a physical barrier that kept slugs out around garden plants.

In this project, you'll test three or more materials as barriers against slugs. The original project used pine wood shavings, oak leaves, and dry, broken egg shells, but you can add additional materials as well. Can any of these materials make a barrier that would prevent common garden slugs from entering and damaging your plants?

Materials list

- Slugs (You can either catch your own or purchase them from a scientific supply house.)
- Two empty, clean two-liter plastic soda bottles
- Large bag of potting soil
- Two yards of fine netting (such as cheesecloth or pantyhose)
- Duct tape
- Handful of small rocks (roughly ¼ inch in diameter)
- Handful of small sticks (roughly 1½ inches to 2 inches long)
- Plywood (¼ inch thick, 13½ inches × 13¾ inches)
- Board (1 inch thick, 3½ inches wide by 6 feet 1½ inches long)
- Box of nails
- Saw
- Scissors
- Hammer
- Measuring tape
- Fresh lettuce (favorite food for slugs)
- At least three barrier materials you want to test: oak leaves, pine wood shavings, dry broken eggshells, or others

Procedures

There are three steps to this project. In the first step, you'll make a slug cage to hold the slugs while they are not participating in the experiment. Then you'll make the experimental test box in which the actual tests will be performed, and finally you'll run the experiments.

If you catch your own slugs, try to find eight slugs of varying size so they can be easily identified. If you order slugs, request that they be different sizes. To make the slug cages, start with two 2-liter plastic soda bottles. Wash them well and replace the lids. Using scissors, cut a rectangular hole in one side. (See Fig. 9-1.) Add a two-inch layer of moist soil on the bottom (actually along the side) and something for the slugs to hide under, such as a piece of tree bark. Put in some lettuce.

After both bottles are finished, put the four smallest slugs in one bottle labelled "Small 1–4," and put the four largest slugs in the other bottle labelled "Large 1–4." The only way you will be able to identify each slug is by its relative size and the cage it comes from. Tape a piece of fine netting over the rectangular opening. Be sure it is securely taped around the edges, or the slugs will crawl out. Leave the box alone overnight to allow the slugs to acclimate to their new home. (See Fig. 9-2.)

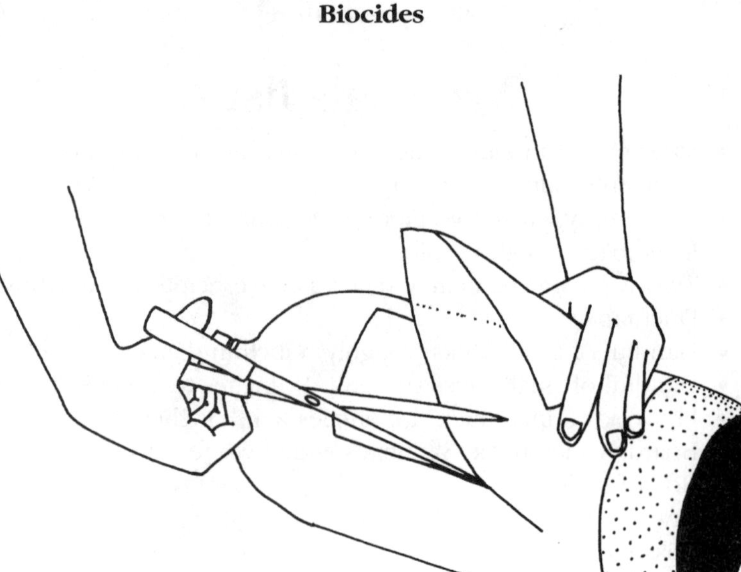

9-1 *Cut a rectangular window out of the plastic bottle.*

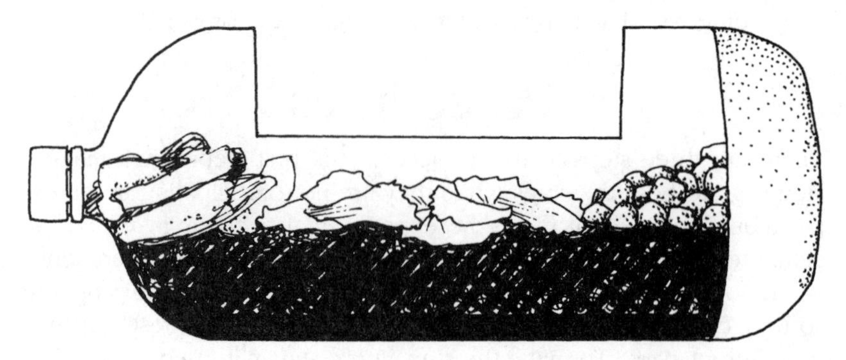

9-2 *This bottle will act as a temporary home for the slugs during the tests.*

Next, you'll make the experiment box. Cut the board (described in the materials list) into seven smaller pieces as follows: five boards should be 12 feet, and two boards should be 13¾ feet. Create a frame from these seven pieces. (See Fig. 9-3.)

With the frame completed, cut the piece of plywood to the correct size as described in the materials list. Lay the plywood on top of the frame and nail it in place. This will act as the bottom of the box. Turn the box over so the open end is face up. Add a layer of moist potting soil, about 2 inches thick, to cover the bottom of all four sections. The soil must be kept damp throughout the project.

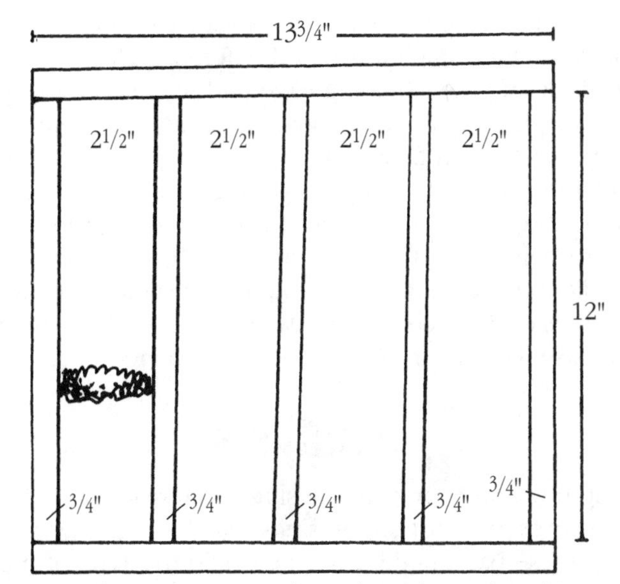

9-3 *The seven pieces of wood are framed as shown here.*

You will place each of the first three materials to be tested in each of the first three sections in the box. (The fourth section will be the control without a barrier.) Place a wall of barrier material across the center of each section about 2 inches thick. If you are using crushed egg shells or any other material that might be moist, air dry them or place them in an oven until they dry before use.

After the barriers are in place, you must prevent the slugs from crawling up the sides of the walls and over the barriers. To do this, fold a piece of duct tape so the sticky part is outside all the way around. Make the piece of tape the correct size to fit on the side wall above the barrier. You'll need two pieces of tape, one for each side wall. (In a real garden, this would not be necessary since there would be no side walls.) Leave one section without a barrier for a control. When the barriers are all completed, place a small pile of lettuce at one end of each section to attract the slug and put one slug at the other end.

You are now ready to begin the experiment. Each slug will be tested in each section twice. Take the large slug ("Large 1") from the large holding cage and place it in the first section on the opposite side of the barrier from the food. Then place "Large 2" in the next section, "Large 3" in the next, and "Large 4" in the control section. Cover the box with the netting and carefully tape the netting to the sides of the box all the way around. Take detailed notes to keep track of which slug is where.

The next day record whether the slugs crossed the barrier to get the lettuce. Use the slime trails to tell where the slugs had gone. Put the slugs back into their holding cage and feed them. Clean the experiment box thoroughly to be sure no slime trails remain and replace each section with fresh lettuce.

Alternate the slugs from each holding cage each day. Since you are alternating days, each slug gets a day (and night) of rest in his cage before being tested again. During each test, place the largest slug in a different section. When you are finished, each slug should have been tested in each section twice. After all the tests are completed, release the slugs near the place you found them.

Analysis

Did any of the barriers prevent the slugs from reaching the food? Did the same barrier work for all the slugs? Were some slugs more adept at crossing these barriers than others? Could this type of mechanical pest control be used in your garden?

Going further

If one of the barriers appeared to be effective, perform a small scale field test. Use a small plot of land, such as your garden, for the test. Prepare enough of the barrier material to surround a section of the garden. Leave another section with no barrier as a control. Observe differences, if any, between the control section and the protected section.

Suggested research

- Mechanical control is the oldest form of pest management. Research mechanical control measures as they were used in the past and as they are being used today.
- Most people think of insects when they think of pests. Slugs, however, are mollusks. Look into the damage caused by non-insect pests. How big a problem do they pose?

10

Biological control

Studying the egg-laying behavior of a beneficial insect

Throughout the world, and especially in less developed nations, a great deal of the food we grow never makes it into our mouths, because insects eat it first. Farmers continually battle insect populations to grow their crops. They use a number of weapons in this fight, but primarily they use synthetic chemical pesticides.

Since Rachel Carson wrote her famous book *Silent Spring* (see page 54), we have realized that pesticides kill more than just pests. They often wreak havoc with entire ecosystems. Many insect species can be successfully used to help control insect pests. The use of such beneficial insects and other organisms to control pests is called biological control, and it has become a state-of-the-art method of fighting insect pests without using chemicals.

Reduced use of pesticides also means less chance of the consumer ingesting these substances, because they often remain on the foods as residues. As our environmental awareness about pesticides increases, the use of biological control agents becomes more popular.

There are three kinds of biological control agents now being used to control insect pests. *Predators*—such as the praying mantis and ladybug—attack, kill, and eat other insects. *Parasites* live on another organism, called a host, and feed on it, but they don't usually kill the organism. Parasites might live on more than one host during their lifetime. Mites, for example, often parasitize insects. *Parasitoids* feed on only one host their entire life and usually end up killing the host, but

only after the parasitoid has matured. There are many species of tiny parasitoid wasps that kill harmful insects. (See Fig. 10-1.) They are all, however, harmless to humans.

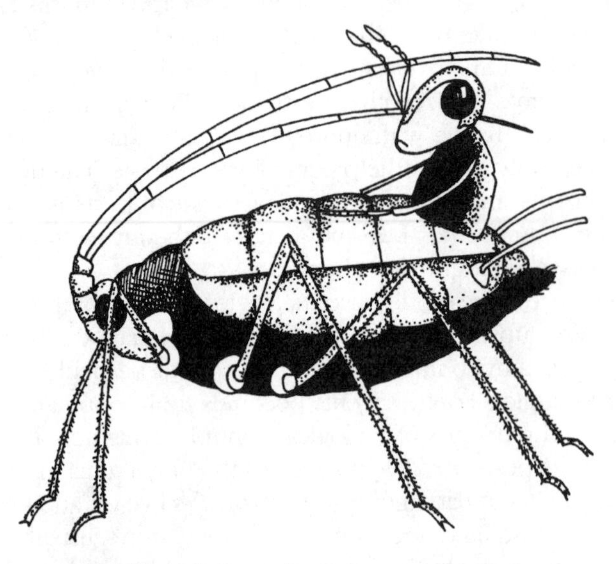

10-1 *Biological control uses organisms such as this parasitoid wasp shown here emerging from an aphid carcass.*

Project overview

This project investigates the life of a parasitoid wasp called *Trichogramma minutens.* (It doesn't have a common name.) This wasp is an egg parasitoid, meaning the adult female wasp lays its eggs in another insect's egg. The immature wasp hatches out of its egg (within the host egg) and eats the inside of the host egg for nourishment. When the immature wasp has become fully grown, it emerges out of the host egg as an adult wasp. The host (the egg) dies. *Trichogramma* is often used to control a common garden pest, the tobacco (also called tomato) hornworm.

This project will investigate the females' egg laying behavior. This is important in determining if a beneficial insect will succeed in controlling a pest. There are two sections in this project. In the first, you'll investigate the parasitoids egg-laying behavior, and in the second you'll investigate the reproductive potential of these insects. Both must be done before a biological control agent can be used.

Studies have shown there is a direct relationship between the size of a female insect and the number of eggs she can lay. Since these wasps are very small, they have a limited number of eggs they can produce, and they must be selective about where they lay their eggs. Laying eggs in the wrong kind of host, a sick host, or a host that has already been parasitized means their eggs have less of a chance for survival. If the female parasitoid is to avoid wasting eggs, she must be able to sense the condition of a host to determine the chances for the survival of her offspring.

In the first part of this experiment, you will study how a mother parasitoid wasp senses a host. How does she determine where to lay eggs? Does her selection criteria change under crowded conditions?

In the second part of this project, you will perform an experiment to determine how effective this biological control agent can be in controlling the tobacco hornworm pest. How long does it take for the adult parasitoid wasp to emerge from within the egg? How many wasps can emerge from one parasitized egg? How many eggs can a single wasp parasitize? What percentage of all the pests are killed?

Materials list

These wasps are visible to the naked eye, but they are very small and require most of the work to be done under a dissecting microscope or a high quality magnifying lens. For Part one, you will need the following:

- At least 50 newly emerged, female *Trichogramma minutens* wasps (These can be purchased from a scientific supply house or possibly from an organic gardening nursery; sexing these insects is explained in the Procedures section.)
- 50 tobacco (also called tomato) hornworm eggs to use as a host for the wasp eggs (available from a scientific supply house.)
- A few petri dishes (or a dish that is ½ inch deep and 4 inches in diameter and has a top)
- Fine pin
- Forceps
- Marker
- Small paintbrush (like those used for painting model airplanes)
- Dissecting microscope or a high quality magnifying glass

For Part two, you will need the following:

- Ten Tobacco hornworm eggs already parasitized by *Trichogramma minutens* (available from a scientific supply house or possibly an organic gardening nursery)
- Ten deep saucers (about ½ inch high and 4 inches in diameter) that can be closed or petri dishes with tops
- Fine forceps
- Small paint brush (like those used for model airplanes)
- Honey
- Spoon
- Dissecting microscope or high quality magnifying glass

Procedures

Part one

First, you will compare the oviposition (egg-laying) behavior of a parasitoid wasp on "good" versus "bad" host eggs. This part continues to determine how overcrowded conditions changes the insect's oviposition behavior.

Take some hornworm eggs (about 20) and gently squeeze them with the forceps until they become distorted, or put a hole in them with a fine pin. Draw a line with a marker to divide a petri dish into two halves. Label one half "good" and the other "bad." Place the bad (damaged) eggs on one side and an equal number of good (undamaged) eggs on the other side of the petri dish. Keep the good and bad eggs separated in the dish.

Now, add five female *Trichogramma minutens* wasps into the dish. Use the paint brush to pick up and move the wasps. Be very gentle, because they are delicate. Females do not have hairs on their antennae as do males, and they don't move around as much as males. As you place each wasp into the dish, close the top.

Observe their behavior under magnification. Watch how the wasps prepare to lay eggs. How do they walk over the host eggs? Record their movements on each type of egg (good and bad). Record how they touch each type of egg. What body parts do they use to inspect the eggs? Do they lay any of their eggs in the bad host eggs? How much time do they spend on the good eggs and on the bad eggs? Watch the wasps for about one hour.

Continuing this part of the experiment, divide a petri dish into two halves once again with a marker. Once again damage about 15 tobacco hornworm eggs as described earlier. This time place 15 "good" eggs on one side and 15 "bad" eggs on the other side of the line. Place 40 female wasps into the center of the dish and replace the top. Begin your observations of their behavior once again. Make your observations as you did in the first part of the experiment. How does their behavior differ when the female wasps are in overcrowded conditions?

Part two

First, you will determine the number of wasps that emerge out of a single parasitized hornworm egg. Put a single parasitized egg in each of the ten petri dishes. (Parasitized eggs are dark-colored. Don't use green eggs since they aren't parasitized.) Be careful when separating the egg from the egg mass. Use either fine forceps or a fine brush.

Place a small drop of honey in each dish. (The honey drop should be as small as this capital O.) Close the dish and seal it with clear tape so the wasps cannot escape. (If they do escape, they won't hurt anything, but your experiment will be ruined.) Keep the dishes at room temperature. Each day, check the dishes for adult wasps. They are small, so look at the dish with the magnifying glass or a dissecting microscope. Continue to observe them every day for four weeks. When they emerge, count the number of wasps in each dish.

Continue this experiment to find out how many eggs a single wasp can parasitize by using a few of the wasps that emerged in the last part of this experiment. Place five, ten, and fifty unparasitized hornworm eggs into three new petri dishes. (See Fig. 10-2.)

Place a single female wasp in each dish. You can tell when an egg has been parasitized because it will darken and turn black or brown within one to two weeks. Count the number of dark eggs to

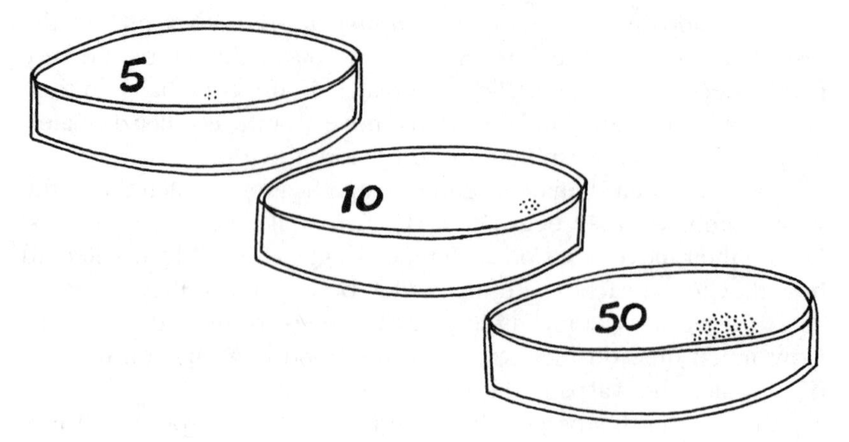

10-2 *Label three petri dishes "5," "10," and "50" and place that number of tobacco hornworm eggs in each dish.*

find out how many eggs a single wasp can parasitize. After the eggs turn dark, allow them to hatch. Once again, see how many wasps emerge out of the eggs in each of the three dishes.

Analysis

From Part one, what parts of the wasp's body are used during the investigation of the host's eggs? Does it appear she can tell a good egg from a bad egg? What happened when there were fewer good eggs and she was in overcrowded conditions? Does overcrowding change the oviposition behavior? How might these results impact the use of these insects to control insect pests?

For the second part, how long did it take before adults emerged from the parasitized eggs? How many wasps can come out of each egg? How many eggs can a single wasp parasitize? Can you estimate how many insect pests can theoretically be destroyed over a given period of time from one female wasp? Draw a chart that plots the population growth curve of a population of parasitoid wasps. How many wasps emerged out of the second set of dishes containing five, ten, and fifty eggs? Calculate how many emerged from each egg for each dish. Chart the results.

Going further

Devise an experiment to determine the specific ratio of ovipositing females to available eggs that results in a change in egg laying behavior. In other words, at what point does overcrowding become an issue to an egg laying female? Or devise an experiment to determine if environmental factors such as temperature affect the wasp's oviposition behavior and her ability to reproduce.

These experiments exclude all mortality factors such as disease, predation, and weather. Devise an experiment that would take mortality into consideration. Consider actual field studies. Try to determine what other specific types of insect pests can be controlled using these two biological control agents. In many cases, biological control is used in an area where chemical insecticides are also sprayed. How does the use of chemical insecticides affect the ability of a biological control agent to kill pests? Can you devise an experiment to test this?

Suggested research

- Research what other factors (besides overcrowding) might play a role in using parasitoids to control pests.
- Read more about other biological control methods that have been used to control insect pests. There are many success stories.
- Contact farmers or distributors that produce "organically grown" fruits and vegetables. Find out what they use for pest control.
- Read about how these and other types of parasitic and parasitoid wasps are used as biological control agents.
- Contact the Entomological Society of America in Lanham, MD.

11

Herbs & germs

Can herbs inhibit microbe growth & control disease spreading?

People try to control harmful organisms in a variety of ways. We use insecticides to control insects and herbicides to control weeds. We fumigate our fruits and vegetables to prevent mold from growing during shipment and while on store shelves. We use antibiotics to control many disease-producing microbes including bacteria and fungi. Pasteurization is used to prevent bacteria from growing in diary products.

We are always looking for new methods of protecting our crops and ourselves from destructive organisms. Recently, we have started to irradiate some foods with radioactive cobalt to prevent microbes from growing. The least toxic solutions for controlling unwanted organisms are always preferred over the use of dangerous chemicals that can cause as much harm as good.

Project overview

There is a saying, "What goes around, comes around." State-of-the-art research is beginning to look into the value of some old folk remedies, including the use of herbal brews. These remedies have been long thought to restore health, prevent illness, and protect against harm. Can herbs be used to prevent the growth of microbes? If so, perhaps they can be safely used to prevent contamination of our foods or even have some medicinal value.

In this project, you'll test to see if certain commonly used herbs are capable of inhibiting the growth of a non-pathenogenic variety of *E. coli* bacteria. You'll use sterilized discs dipped in various herbal brews (much like antibiotic discs are used) to see if they can control the spread of this microbe and possibly control the spread of disease.

Note: This project requires the use of standard microbiology techniques, such as sterilizing glassware, streaking agar plates, and handling bacterial cultures. Your sponsor must be knowledgeable about these techniques and the safe use and disposal of these cultures. For more information about the techniques, see *Microbiology: High-School Science Fair Experiments* by McGraw-Hill (#0156646).

Materials list

- Safety goggles and laboratory gloves
- Small pot for boiling
- Measuring cup
- *E. coli* (a non-pathenogenic strain available from a scientific supply house)
- Twenty-one pre-packaged nutrient agar plates (You can prepare your own with nutrient agar mix and petri dishes—all available from a scientific supply house.)
- Sterile inoculating loops
- Bunsen burner
- Incubator
- Forceps
- Two beakers for each herb used (used to boil and hold the brews)
- Glass jars with tight screw caps that can hold one cup for each herb listed below
- Sterile discs (available from a scientific supply house)
- Autoclave to sterilize glassware and utensils (optional)
- At least ½ cup of the following herbs, which can be fresh or dried (you can add your own choices): basil, borage, chamomile, chickweed, garlic, peppermint, and savory

Procedures

Follow all safety procedures as recommended by your sponsor throughout this project. Wear safety goggles and laboratory gloves.

To make the herbal brews needed for this project, follow these instructions. Mix ½ cup of each herb with one cup of water in a small pot. Bring the mixture to a boil for five minutes, then let stand for 24 hours. Prepare one sterile jar for each herb by using an autoclave or by boiling the jars using a sterile technique. After 24 hours, pour the liquid off the solid into another beaker and reheat this liquid once again to a boil. Immediately, pour the liquid into the sterilized jar. Seal tightly. Be sure to label each jar with the proper herb brew. (See Fig. 11-1.) The herbal brews are now ready.

11-1 *Each jar will contain an herbal brew which will be tested for its antimicrobial qualities.*

For the next part of the project, you will prepare the agar plates for the bacteria and place herbal brew discs on the plates to see if they can inhibit the growth of the bacteria. Begin by preparing 21 nutrient agar plates or purchase pre-mixed, sterile nutrient agar plates. Using proper aseptic technique, use an inoculating loop to streak each plate thoroughly and consistently with the *E. coli.* Check with your sponsor or teacher about the proper streaking technique to use.

After all the plates are inoculated with *E. coli*, use a marker to divide the top of each plate into quadrants. (See Fig. 11-2.) Next, dip a sterile disc into one of the herbal brews using a sterile forceps. Gen-

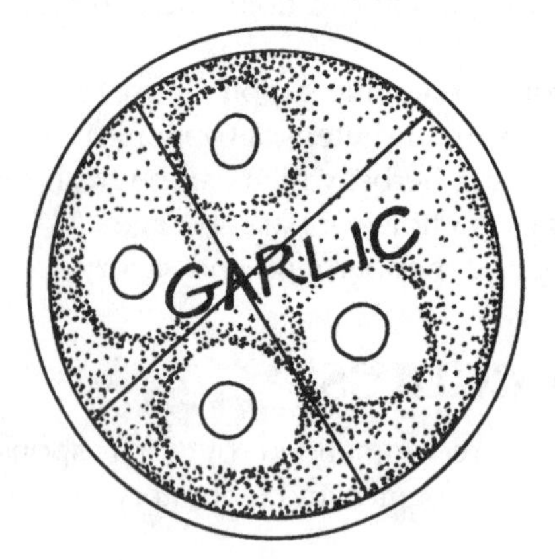

11-2
Look for halos around each disk indicating microbial inhibition.

tly place the soaked disc onto one of the quadrants in the properly labelled agar plate. Soak three more discs with the same brew, and place them in each of the three remaining quadrants of the same plate. Each plate will have four discs. Label the plate with the name of the herbal brew as shown in the figure. Repeat this procedure for all the herbal brews.

After all the plates are ready, place them in an incubator at 37 degrees Centigrade for 24 hours. After this time remove the plates and note the growth of the bacteria in each plate making special note of the growth or lack of growth around the discs.

Analysis

Look for halos of no growth around the discs. These halos are areas where the herbal brew, which is soaked into each disc, inhibited the growth of the bacteria. The larger the halo, the greater the inhibition. Measure the width of the halos around each disc. Take the average for each plate (all four discs), and then compare the averages of all the plates. Did any of the herbal brews effectively inhibit growth? Fill in a table. (See Fig. 11-3.)

Going further

Did more than one herb successfully prevent growth? Investigate what chemicals are found in any of the herbs that prevented growth to determine the chemical responsible for the inhibition. Consider doing thin-layer chromatography to separate out the components of the herb, or do a literature search to find the biochemical component of those herbs to determine the active ingredient.

Suggested research

- Research old folklore medicine and try to tie it in with modern day science discoveries related to this project.
- Look into natural extracts that have been used in the past and today to cure disease and prevent illness. What does this subject have to do with scientists concerns over the loss of biodiversity on our planet and the rapid rate at which plants and animals are becoming extinct?

HERBAL BREW	AVG. DIAMETER HALO AROUND DISK					NOTES
	#1	#2	#3	#4	AVG. FOR ALL	
1. PEPPERMINT						
2. GARLIC						
3.						
4.						
5.						
6.						
7.						

11-3 *Create a table like this one to compare the degree of antimicrobial effectiveness of each herbal brew.*

12

Slug attractants

What substances can attract & kill slugs?

Many gardeners know that a cup of beer left around the garden is a good way to attract slugs and keep them from munching on the greens. This mechanical means of pest control works quite well in many parts of the country. It reduces the need for chemical pesticides that are dangerous to people and the environment. Learning more about how old pest control measures work can lead to new, more advanced methods of pest control.

Slugs are a common garden pest in all parts of the country (see page 66). Slugs are classified as gastropods, which are mollusks. This means they are more closely related to snails and clams than worms, as many people believe.

Project overview

What is in the beer that attracts slugs and similar pests? Is it the alcohol, the yeast, the hops, or simply the water? In this project, you will determine what it is about beer that attracts slugs. In this project, you'll find which type of beer slugs prefer. You'll compare the slug-attracting qualities of beer with other substances, including plain water, yeast, and alcohol.

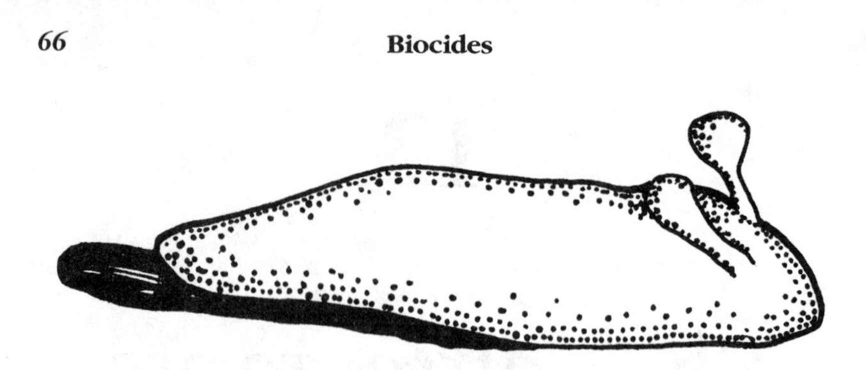

Materials list

- About 15 slugs (caught or purchased from a scientific supply house)
- A selection of beers, including one light beer, one pilsner, one dark beer, and one ale (Your sponsor or parent will purchase the beers and monitor their use throughout the experiment.)
- Brewers or bakers yeast
- Grain alcohol
- About five shallow, glass bowls
- A large metal or plastic tub in which the slugs and the bowls containing various test substances can be placed
- Marker to write on glass

Procedures

Label each bowl with the name of one of the beers. Place about one inch of each beer in each properly labelled bowl. Place the bowls in a straight line in the bathtub or other tub large enough to accommodate the bowls and slugs. (See Fig. 12-1.) Place five slugs of varying sizes (so you can identify which is which) in a horizontal line facing the bowls. Allow enough time for the slugs to choose a beer. Record which slug preferred which beer.

Repeat this experiment four more times, but rearrange the beer each time so they are in a different order. Release the slugs from the same location to assure that the chances of a slug finding a certain bowl are random.

After completing these tests, run a similar series with the next five slugs and then finally with the last five slugs. Record all the results, and determine which beer was favored by the slugs. Even if there is

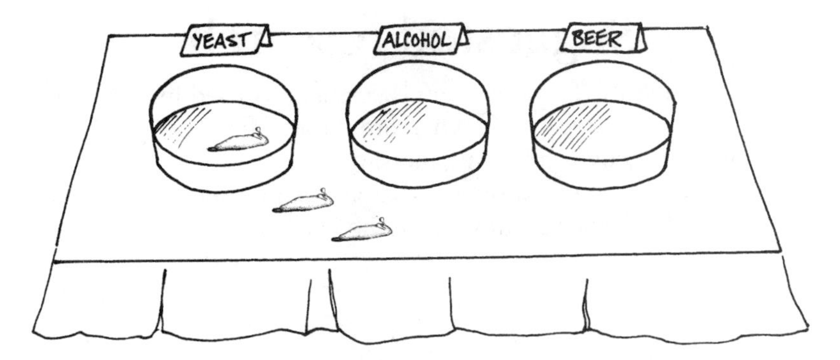

12-1 *See which dish attracts the most slugs.*

no significant difference between some of the beers, choose the beer most often selected by the slugs.

Now you'll compare the slugs' favorite beer to specific components of the beer. What is it about the beer that attracts the slugs? Is it the entire brew (meaning the beer) or some component of the brewing process such as the yeast used or the end-product alcohol?

To proceed, once again set up four bowls, but this time pour one inch of water and add a few tablespoons of brewers yeast into one bowl, pour one inch of grain alcohol into another bowl, pour one inch of the favorite beer (selected in the first part of the project) into the third bowl, and pour just one inch of water in the last bowl. Repeat the same series of tests as you did in the first part of this experiment. Record the slug preferences for each series of tests.

Analysis

Did the slugs prefer the water, yeast, alcohol, or beer? Does it appear that it is the beer or some component of the beer that attracts the slugs? If it is a component of the beer, which one?

Going further

Select the component preferred by the slugs and continue the experiment by looking for the preferred concentration of that substance. Fine-tune the experiment to find the favorite slug attractant. Once you find this substance, run some small scale field tests in your garden to see if it works in nature as well as in the lab.

Suggested research

- Research the fermentation process as it is used in brewing. What is produced and when are these substances produced? What exactly is attracting the slugs?
- Look into other substances that are being used to control slugs and other gastropods in the garden.

Part 4

The lives of animals

The biodiversity of our planet is amazing. With over one million different types of animals on our planet, the research possibilities are virtually endless. This section provides projects that use different groups of animals to allow you to investigate the interesting aspects of their lives.

The first project in this part investigates the hissing sounds of a large cockroach. Are all hisses alike, or do those made during aggression differ from those made when alarmed or during mating?

Anoles, also called American chameleons, change color from green to brown or yellow. Do large anoles change color as quickly as smaller anoles? This second project investigates whether size has anything to do with the speed of color change in anoles.

In the third project, you will discover whether or not it's possible to communicate with fireflies.

Mosquitoes are the subjects of the fourth project. How does temperature affect the growth and development of these insects?

Many of us take vitamins to supplement our diets. If vitamins are so important to our growth and continued health, could they be beneficial in the regeneration capabilities of a planaria? The last project in this part investigates whether a supplement of a vitamin enhances a planarian ability to regenerate lost parts.

13

Madagascar hissing cockroaches

Are all their hisses alike?

Many organisms communicate vocally with members of the same species and of other species. They vocalize to attract a mate, warn others of danger, and intimidate others. Birds are probably the most noticeable animals that communicate through sound.

Many species of insects use vocalization, also. On a warm summer night this vocalization is quite obvious when you hear crickets, katydids, cicadas, and a multitude of other insects producing a symphony of mating calls.

Project overview

The Madagascar hissing cockroach has been given its common name not only because it originally comes from the island of Madagascar, but also because it produces a noticeable hissing sound. These large insects hiss during courtship, mating, aggression, alarm, high exertion, high agitation, and flight.

The hissing sound is produced when air is expelled from a pair of specialized spiracles, which are the organs used to transfer air in and out of the insect's respiratory system. This system delivers oxy-

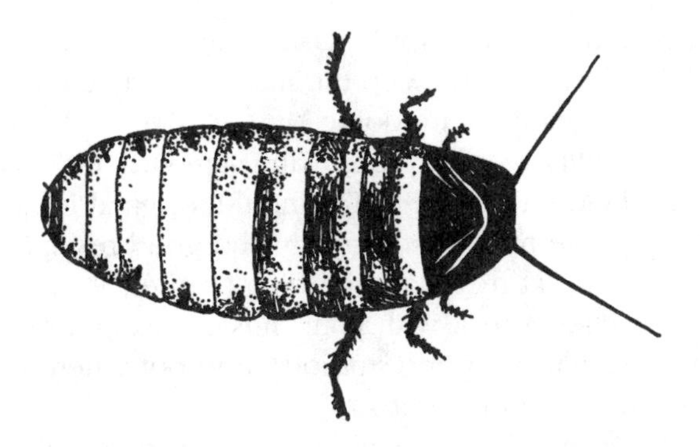

gen and removes carbon dioxide throughout the body. (Insects don't have lungs.) The hissing sounds occur within a broad frequency range between 2 kHz and 20 kHz and with an amplitude from about 40 dB to more than 80 dB.

Hissing is primarily performed by the male of the species. The only sounds produced by females occasionally occur during mating, but they do not hiss during any of the other activities just mentioned.

This project attempts to determine if there are unique types of sounds produced for each type of behavior. For example, is the hiss made to ward off others different than the hiss made during courtship, or is a hiss simply a hiss regardless of the behavior?

Materials list

- Twenty to thirty Madagascar hissing cockroaches, *Gromphadorhina portentosa* (Many universities have colonies of these interesting insects. If possible, obtain permission to use an existing colony or to borrow some specimens. They are also available at scientific supply houses or through insect breeders.)
- Aquarium or terrarium to house the cockroach colony (alternatively, wood and glass enclosures can be used)
- Glass or plastic tank cover to raise the temperature in the tank
- Ground cover for the tank (can be sand, gravel or wood chips)
- Toilet tissue tubes or mailing tubes to act as hiding places for the cockroaches in the tank
- Flower petals, apples, and lettuce for food
- Clean water to be misted onto the walls of the tank twice daily for the cockroaches to drink
- Water mister (to mist the colony with warm water)
- Heating pad, heating rock or lamp to keep the colony environment between 75° and 80° F

- Access to a photographic darkroom to run the experiment (This allows you to keep the insects in the dark while you can see what you are doing.)
- High-quality tape recorder using standard audio cassettes
- Hand-held microphone (During the original project, inexpensive microphones provided good results since they were better at recording the high frequencies produced by the roaches. Speak with your music teacher, or contact a professional recording studio to find out where you can borrow this equipment.)
- Audio scanning device and software that can digitize sound (convert the audio tape into digital form) so it can be printed out as you see in Figs. 13-2 through 13-4. (There are many such devices available. A Macintosh computer with a MacRecorder and SoundEdit Pro was used during the original project. Contact a computer teacher or a professional recording studio to find out where to borrow this equipment.)
- Plenty of patience to wait for behavior patterns to occur naturally
- Small bottles of airplane paints (as many colors as possible) to identify individuals
- Fine model airplane paint brush

Procedures

These insects are most active during the night in the dark and when it is warm and very humid. To meet these requirements, set up the experiment in the photographic darkroom. To increase the temperature in the tank, spray the tank with warm water, place a glass bottle of hot water in the tank and cover the entire tank with a glass or plastic cover. To increase their activity you can place some flower petals in the tank. Try to maintain a temperature of about 23 degrees C. Let the insects become acclimated to their new environment for about 45 minutes while you prepare your audio equipment.

Position the microphone in the tank. Be sure the tape recorder counter is set to zero. (See Fig. 13-1.) When all is ready, remove the glass top, turn on the recorder, and begin to observe. When a specific behavior occurs, such as aggression, mating, or alarm, note the behavior and the counter number. (You will use the counter number to match the sound with the behavior.)

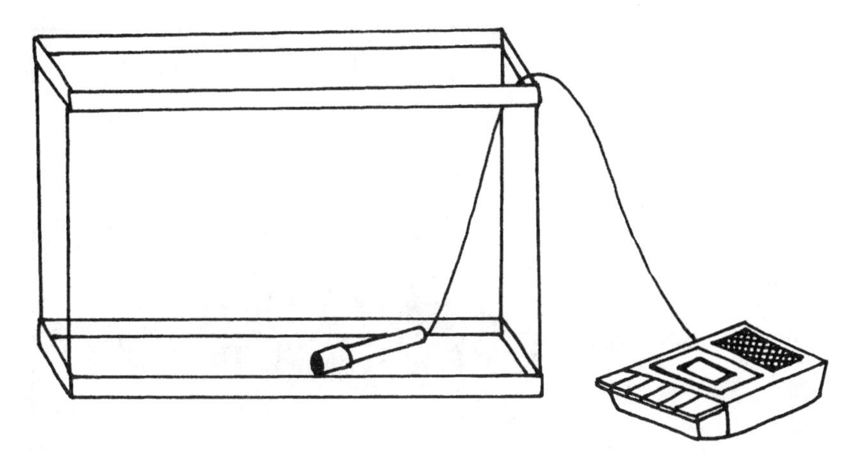

13-1 *Place the microphone in the tank holding the cockroaches.*

During the original project, the hours between 10:30 p.m. and 1:00 a.m. were found to give the best results. You will have to spend many hours observing the insects' behavior. You might want to have a partner assist with these observations.

After each session, scan the audio cassette containing the hissing sounds to locate each sound. Every time a hiss is heard, record the counter number. You can then match the sound, as indicated by the counter number, with the behavior you observed, also indicated by the number that you wrote down earlier. After the sound and the behavior have been matched, you can digitize the sounds.

Now remove the audio cassette from the cassette player, and insert it into the scanning device (for example, MacRecorder), which will digitize the sound. Once the transfer has been made, use the software program (for example, SoundEdit Pro) to generate graphs of the sounds. Create graphs that represent time, amplitude, and frequency. (See Fig. 13-2.)

Analysis

Categorize all the sound graphs by behavior. Analyze the data by looking for similarities within each category and between categories. Do certain behavior patterns, such as mating, produce different sound graphs than aggressive behavior patterns? Do these hissing cockroaches "speak" differently to one another depending on what they have to "say"?

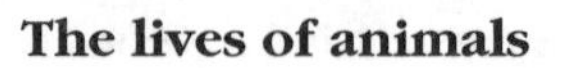

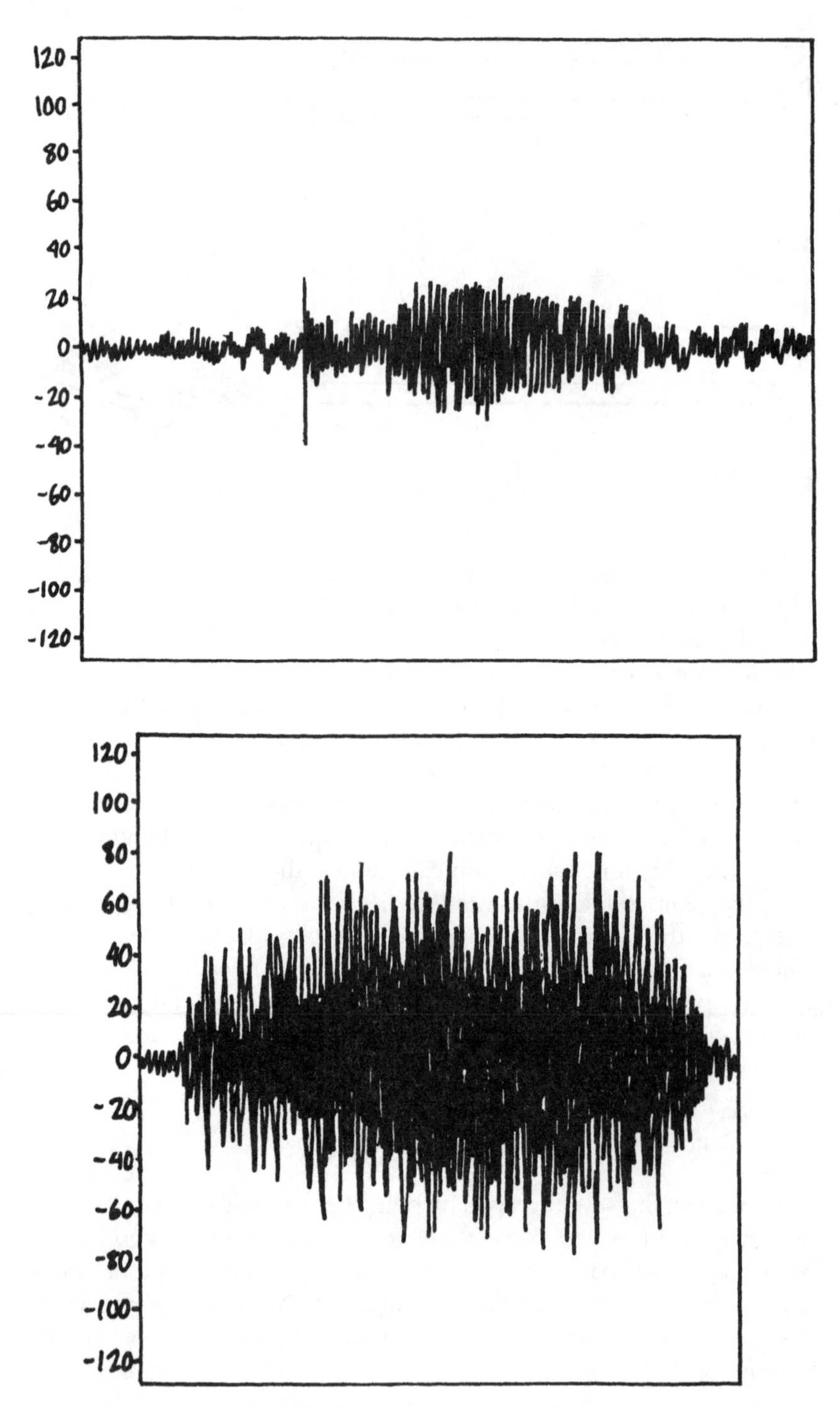

13-2 *Sound graphs clearly depict qualitative differences between sounds.*

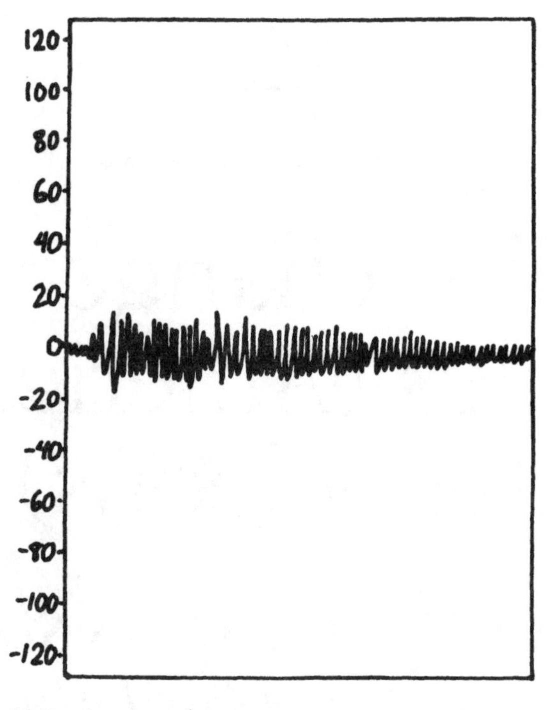

13-2 *Continued.*

Going further

Modify the experiment to determine if different individuals produce similar or different hissing sounds during similar activities. Identify each insect before beginning the project by painting a tiny colored dot on the back of each insect's thorax. When taking notes, record not only the behavior elicited, but also the individual (colored dot) that produced the behavior. Do all individuals produce the same sounds for the same behavior or does every individual speak his or her own language?

Suggested research

- Research insect vocalization. Do any other insects produce sound via their spiracles?
- Study insect communications in general. The literature is voluminous.
- Notice the "eye spots" on these cockroaches. Research the purpose of these spots. Do you think they have a purpose?
- Look into other insects and other animals that use similar mimicry. See the front cover of this book.

14

Color change in anoles (American chameleons)

Does the lizard's size affect the speed with which it changes color?

Green anoles, commonly known as American chameleons, are known for their fascinating ability to change color. The exact cause for this change is still considered a mystery. Recent research has shown that changes are at least partially controlled by the autonomic nervous system and hormones.

Color change appears to be brought on by many factors, including light, temperature, mental state, health, and the time of day. The most rapid changes occur when they become frightened or when there is a rapid change in their environment.

Project overview

The color of an anole can range from shades of green to yellow to gray and brown. Anoles are usually green in darkness or when excited and brown in cold temperatures or during the brightest hours of sunshine. At night anoles are usually pale green with a whitish stomach.

Since so much is still unknown about color changes in these creatures, it would be interesting to investigate some aspect of the color changing process. This project looks into whether the size of an anole affects the time it takes for the color change to occur. Will a large anole change from one color to another more quickly or more slowly than a smaller anole under similar conditions?

Materials list

- At least five anoles of varying sizes (available at pet stores)
- At least a 15-gallon terrarium or aquarium with a mesh cover
- Thermometer
- Rocks, pebbles, and gravel
- Terrarium heating unit, heat rock, or heat lamp
- Flies and mealworms (food for the anoles)
- Water mister bottle
- Stopwatch
- Potted plants that will fit into the tank
- Balance
- Ruler
- Camera and film
- Paint chips (shades of green, brown, and yellow)
- Poster board to create a color wheel
- Scissors
- String
- Clear box about one inch high to temporarily hold the anole while it is being measured (a clear, plastic container will work well)
- Two coffee cans with lids
- Wooden matches

Procedures

First, you must set up a comfortable terrarium for the anoles. Cover the floor of the terrarium (aquarium) with pebbles, gravel, and soil. Add at least one green potted plant, flat rock, and twig. Insert a heating unit or heating rock into the tank or place a heating lamp over the tank. Place the lid over the tank.

Position the thermometer in the tank to monitor the environmental temperature of the terrarium. Place the anoles in the tank. The lizards will require at least two flies each per day. Spray the plants with water daily. The temperature should remain between 20° and

38° Celsius at all times. Check with a pet store operator and/or read literature about proper care for your anoles.

Next, take initial measurements of your anoles. To measure the length of each anole from the tip of the tail to the tip of their nose follow these steps. Cut a piece of string about one foot long. Place the anole in the plastic, transparent container. Since the container is over one inch high, it will keep the anole in an extended (flat) position. Put one end of the string, on the top of the container, over the tip of the anole's tail. Run the string along the contour of the anole's body. Mark the spot on the top of the container where the string reaches the tip of the anole's nose. You'll have to raise the string up slightly in areas where the animal's body is curved upward. Mark the spot on the string where it reaches the tip of the nose. Then remove the string and use the ruler to measure the length of the string from the end to the spot you marked. This represents the length of that anole. After recording the length of this anole, repeat this procedure for each animal.

Number each of the animals from one to six, smallest to largest. After you've recorded their length, take each of the numbered anoles and measure their weight on the balance and record this data. Measure the animals to within 0.1 of a gram.

Before you begin the actual experiment, prepare an anole skin color chart that ranges from light to dark shades of green to brown. Use at least ten colors, and number each of them. (See Fig. 14-1.) Begin the actual experiment by removing all green plants out of the tank so only brown items remain. Gradually, all the anoles will turn brown. When one of the anoles has reached the darkest color brown represented on the chart, record the temperature in the tank, remove the anole from the tank, and begin the stopwatch. Immediately begin to play with the animal. This will cause excitement and cause it to turn green. When the anole reaches the lightest shade of green as represented on the chart, record the time from the stopwatch.

14-1 *Create an anole (American chameleon) color chart depicting the range of colors the lizard turns.*

Be sure to play with the anoles each time in a similar fashion and in a similar environment. Repeat this entire procedure with all five anoles. Repeat this procedure with each anole at least ten times.

Analysis

After all the tests are complete, average the data for each anole. Create a bar chart for each anole. (See Fig. 14-2.) The temperature should remain constant for each trial. If it does not, it must be considered a variable. After studying the data for each anole, compare all the data by creating a chart plotting all the anoles' data. (See Fig. 14-3.) Plot the data in sequence from the smallest (#1) to the largest anole (#5). Does there appear to be a correlation between the size of the anole and the rate of color change?

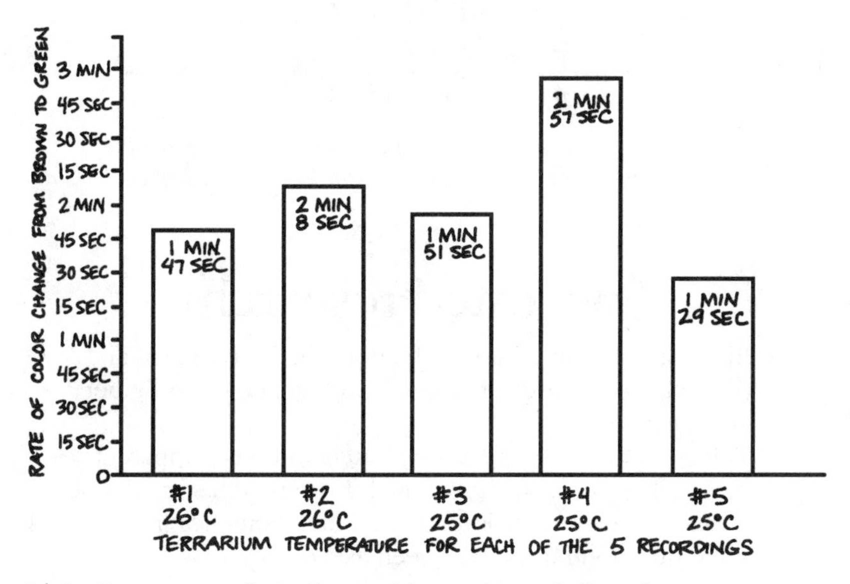

14-2 *Create a graph similar to this one for each lizard.*

Going further

This project tests the speed of changing from brown to green. Continue the project by seeing if you get the same results with the anoles going from green back to brown. See if temperature affects the results or if the cause of the color change (i.e., fright vs. background change) affects the results, as well.

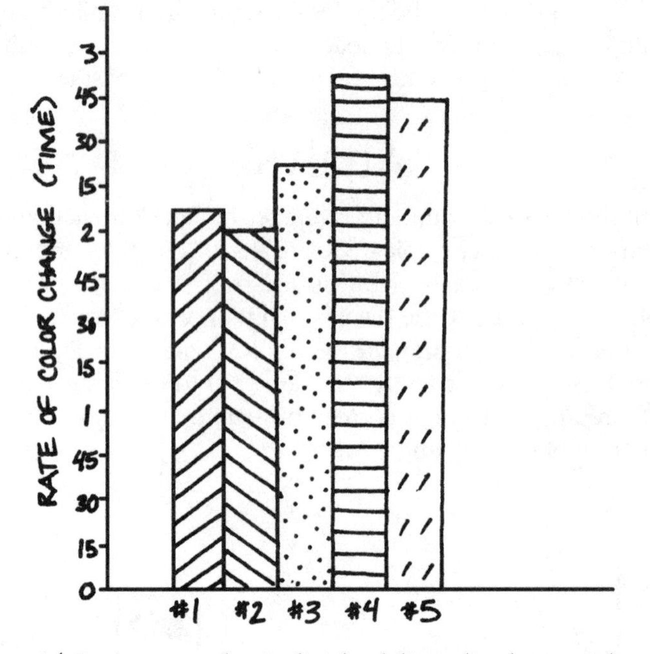

14-3 *Average the individual lizard's data and compile it into one comparison chart similar to this one.*

Suggested research

- Investigate the latest research about the biochemical and physiological causes of color change in anoles and other animals.
- This is an example of basic research with little apparent practical application. How might this research someday serve a useful purpose? Look for some other examples in which the same question might have been asked.

15

Communicating with fireflies

Can you understand the "visual" language of fireflies?

Bioluminescence (light produced by living organisms) is used by many animals. It is most common in organisms that live in the deep, very dark oceans where finding a mate might be impossible without the help of these guiding lights. There are terrestrial animals, however, that also use bioluminescence.

Fireflies use bioluminescence to locate a mate by a certain sequence of flashes. Different species use a different type of flashing sequence. In most species, the female remains on the ground flashing its light to attract males. (See Fig. 15-1.) When males of the same species see the flashes they respond with their own flashes until the two meet.

Project overview

Different flash sequences allow individuals to locate only members of their own species. These sequences are crucial to the successful union of members of the same species and to the continuation of the species. In some species, the shape of the light attracts mates. In others, it is the sequence of the flashing lights that attracts mates. Some firefly predators have evolved bioluminescence to attract a would-be mate, which they instead devour.

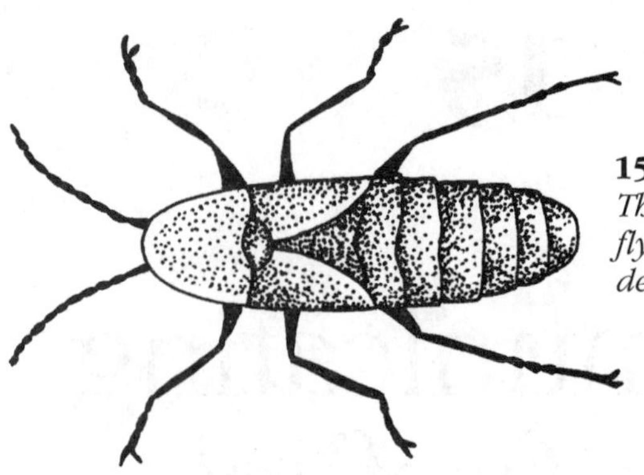

15-1
The female firefly cannot fly since it has poorly developed wings.

In this project, you will first observe fireflies communicating in the wild, and try to identify different sequences of flashing being used. In the second part, you will catch some of the fireflies you've been observing, and try to communicate with them by imitating these light flashes. Is it possible to get fireflies to respond to artificial flashing lights (a flashlight) by imitating their flashing sequence? Can you determine their flashing sequence by their response or lack of it?

Materials list

- An area where fireflies are often seen
- Jar with a lid or some cheesecloth to cover the jar and rubber bands to hold the cover in place
- Small narrow flashlight (preferably a penlight)

Procedures

Start looking for firefly lights at dusk during a warm summer night. Look in an area where you have previously seen fireflies. Bring a flashlight along so you can see where you're going and so you will be able to take notes. Observe and record the movement patterns of flying fireflies as they produce light. (Since they are flying around, they will be males.) Does the light move in a straight line? Does the light curve up or curve down? Does it swoop up and down?

The flight pattern not only refers to the movement of flight, but also the number and sequence of the flashes. Try to count the number of flashes and the flash sequence created during the flight pattern. Take notes and make sketches of the flight patterns you observe. (Include the time these observations are made.) Are the patterns consistent among many individuals? Can you identify more than one pattern (meaning you are probably seeing more than one species)?

Continue your observations for about one hour. Watch them for a 20-minute interval, take a break, and then return to observe again. You might want to try a few different locations if you don't have success at the first site.

After you have studied the flashing behavior in the field, begin the second part of the project. Collect about ten of these fireflies, and place them in a jar. Put the lid on the jar loosely or cover it with cheesecloth and hold it in place with rubber bands. Bring the jar into a dark room. Observe the lighting patterns and record your observations. (If you must turn on the flashlight to write, do this under the table so it doesn't interfere with the project.)

Remove one of the insects, and place it in a separate jar. Place the jar containing the other insects in another room so they don't interfere. Hold the flashlight about one foot from the jar and flash the light in the sequence you identified in the field. Repeat this sequence for 60 seconds. Notice if the insect appears to respond (flash back to you) in any kind of pattern. Record your observations during periods when you flash the flashlight and again when you do not flash. Repeat this procedure for the other nine fireflies and then analyze the data.

Analysis

In the first part of the project, did you detect a definite sequence used by one particular species of firefly? Could you tell how many different species of firefly you observed in the field by the number of flashing sequences? When you mimicked a female firefly with the flashlight, did the fireflies respond to your flashing? If so, did their sequence respond to yours? Do all the fireflies elicit the same flashing sequence in response to yours? Is it possible to communicate with fireflies?

Going further

Devise an experiment to investigate whether the males and females of the same species use the same or a different flashing sequence. Remember the females usually remain on the ground, and the males fly around.

Suggested research

- Read about "cold light." Why is a firefly's light considered the perfect light? Compare the light produced by an incandescent bulb and a firefly's light. (See Fig. 15-2.) See if you can find any products that use a synthetic version of this cold light.
- Study the mating behavior of fireflies.
- Research how bioluminescence is used in other animals.
- Research how bioluminescence works.

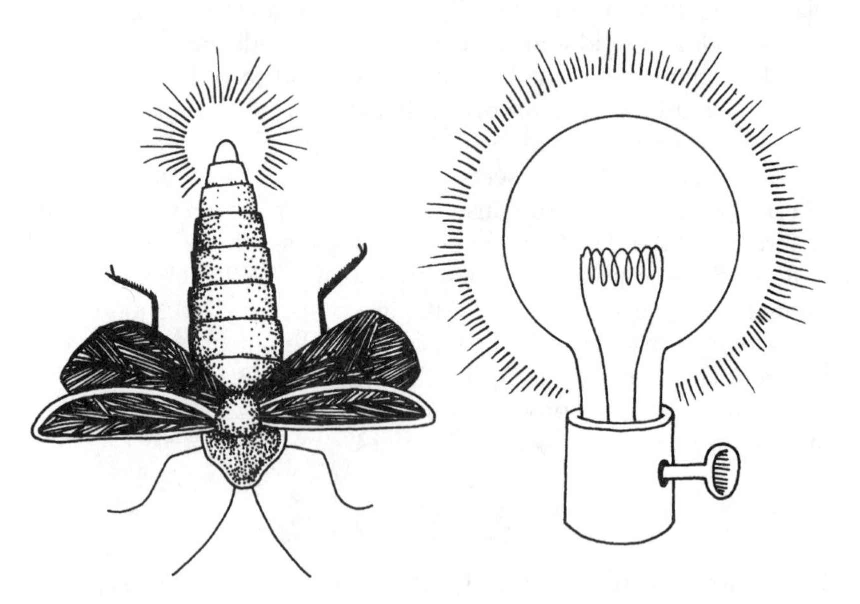

15-2 *Study the differences between the firefly's light and a typical incandescent light bulb. (Art not drawn to scale.)*

16

Temperature & mosquitoes

How does temperature affect the development of mosquito larvae?

Environmental factors, such as temperature, availability of water, and the chemical composition of the soil or air, all play a vital role in determining the fate of organisms. Some environmental factors play roles that we don't usually consider. How does temperature affect the lives of mosquitoes? Does temperature determine whether they live or die? Does it determine whether they grow large or remain small? Does it determine whether their young will develop into males or females?

Project overview

This project investigates the effect of temperature on the development of mosquito larvae. There are two parts to this project. In the first part, you will see if temperature can kill the mosquito at certain stages of development. In the second part, you'll see if temperature affects mosquito development in an unusual way by determining if it affects sex ratios.

Part one

Harsh temperatures can kill any organism, plant, or animal. Many insects have evolved a unique method of surviving through harsh times. Those insects that develop through complete metamorphosis change their form many times throughout their development. The most common example, of course, is the butterfly which passes through the egg, caterpillar, and chrysalis stages before becoming an adult butterfly.

Some insects living in climates that get quite hot or cold during certain times of year survive the harsh seasons by timing their growth so that the stage best suited for survival occurs during the harshest season. In the cold Northeast, this is probably the egg or the pupae, which is enclosed in a tough encasement.

Part two

The number of males to females produced in a population is called the sex ratio. In most populations the ratio is close to 50:50, meaning for every 100 offspring there will be 50 males and 50 females. There are many exceptions to this rule, however. In some instances the sex ratio varies. This might be due to changes in season, predator-prey relationships, changes in weather and so on.

Materials list

- At least 180 mosquito larvae, often called *wrigglers* (These can easily be found during the warm months in stagnant bodies of water, or the eggs can be purchased from a scientific supply house)
- Refrigerator with a freezer
- Incubator (or a warm place in your home or school that maintains a temperature of about 90° F, such as a boiler room)
- Aquarium fish food (available at pet stores)
- Eye dropper
- Ten similar, quart-sized jars with screw caps
- Enough cheese cloth to make covers for all the jars
- Large rubber bands to hold the cloth on the mouths of the jars
- Small aquarium fish net to collect the wrigglers
- Paint brush (like the ones used for painting model airplanes)
- Killing jar (You can make one with a wide mouth jar about the size of a jelly jar, a screw cap, a cotton ball, and some

nail polish remover, or you can purchase one from a scientific supply house.)

- Dissecting microscope or magnifying glass to sex the adult mosquitoes

Procedures

Mosquito larvae, called wrigglers or eggs (usually found in groups or rafts), can be found in stagnant waters, such as a small pond, a bird watering tray, or long standing puddle. (See Fig. 16-1.) You can make your own still body of water by simply leaving a bucket of water outside during the hot summer days.

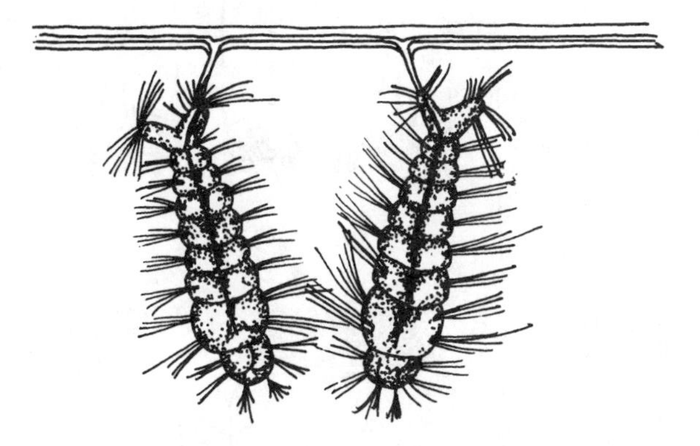

16-1 *The mosquito larvae depends on a special breathing tube that allows air to enter its tracheal system.*

Collect as many wrigglers or egg rafts as possible. They can usually be found in large numbers. You'll need at least 180 for this experiment. Place all the wrigglers (or eggs) in one of the jars filled with water, and label it "mosquitoes." (Use the same water where you found the mosquitoes.) After collecting the mosquito young, fill some extra quart-sized jars with water to be used later.

Distribute the extra water so there are equal amounts of the water in each of the nine jars. They need not be filled, but they should have at least three inches of water. Then divide the 180 wrigglers from the jar marked "mosquitoes" into the nine other jars so they each have at least 20 mosquito young. Cut out a piece of cheese cloth for the top of each jar, and use rubber bands to hold the cheese cloth tightly in place. The wrigglers cannot fly, but when they pupate and emerge as adults, they will be able to fly. Keep the water in the jar marked "mosquito" (now empty of mosquitoes) in case you need additional water later in the experiment.

Label three of the jars "cold," three "warm," and three "normal." (See Fig. 16-2.) Place the "normal" jars in a room that maintains a temperature of about 20° C (roughly 70° F). Place the jars labelled "warm" in an incubator at 35° C (95° F). Place the "normal" jars in a room that maintains normal room temperature. Try to keep all conditions, other than the temperature, the same for all jars in all three groups. For example, keep them all in the dark.

16-2 *You'll have three sets of three jars, including three "WARM," three "COLD," and three "NORMAL."*

Each day observe all the jars and record what you find. Look for wrigglers that have pupated and those that have died. Record all the information daily. Continue the experiment for a few days until you notice adults flying around under the cheese cloth in some of the jars. As this happens, take the jar and place it in the refrigerator for about five minutes to slow the insects down. You can then use a small brush to move them into the killing jar.

To prepare the killing jar, soak a cotton ball with nail polish remover, and place the ball in the jar. Once ready, transfer the mosquitoes to the killing jar. Leave the mosquitoes in the jar for about ten minutes. Once the mosquitoes are dead, place them in a Petri or similar dish, and place them under the dissecting scope or look at them with a magnifying glass.

Sex each individual by first looking at their compound eyes. In most species, the male's eyes are much closer together than the females. (See Fig. 16-3.) The male antennae are usually bushier than the females.

Count the number of males and females for all nine jars after all the adults emerge. Then record the data and calculate the averages for each group of three.

16-3 *You'll have to sex the mosquito adults to determine the ratio of males to females.*

Analysis

First, compare the number of deaths in each jar. Did any of the three temperature groups prove partially or totally fatal to the mosquito larvae? If so, does this mean the mosquito could not survive in regions that attained these temperatures? How might the mosquito survive these temperatures in nature?

For those mosquitoes that did survive and reach maturity, how did the sex ratios differ, if at all? Did any of the jars contain more females or more males? Was the difference significant? Does temperature affect the mosquitoes development resulting in different sex ratios?

Going further

Continue the first part of this project to see if the larvae or pupae that did not mature to adults are able to continue their development when removed from the temperature extreme (cold or hot) and placed in room temperature. To do this, take the cups containing the non-developing larvae or pupae and place them at room temperature for a few days to see what happens. Did any of the "dead" larvae continue to develop and turn into adults? How is this significant in nature?

In the second part of the project, did one of the temperature groups produce different sex ratios? If so, repeat the experiment, but push the temperature until it becomes a limiting factor. For example, if high temperature caused a change in sex ratio, repeat the experiment, but increase the temperature even higher until one of the experimental jars results in the death of all the larvae. How did the sex ratio change until the temperature became a limiting factor? Did you end up with all males or all females?

Suggested research

- Study how temperature affects the development of other organisms and find out why.
- Read about the development of amphibians.
- Research more about how complete metamorphosis protects an insect's development from temperature extremes.

17

Vitamins & regeneration

Do vitamins affect the regeneration of planaria?

Regeneration is an ability found, to some extent, in all organisms. Plants can grow back stems, leaves, and flowers, as long as their roots aren't destroyed. Earthworms can grow back lost heads. Starfish and sponges can grow back lost parts. *Planaria*, a genus of common freshwater flatworms, have the ability to regenerate back into whole worms from almost any piece. (See Fig. 17-1.)

In humans, our epidermal skin cells are continually discarded and replaced. Wounds heal, and broken bones grow back together, but we cannot grow back lost legs or arms.

Project overview

Vitamins are associated with maintaining proper growth and health and with the basic functioning of life itself. Vitamin C has been called a "safe, over-the-counter healing product" by many sources, and studies have proven it to have a positive effect on healing. Vitamins E and B-12 are known to play some role in forming muscles and tissues in humans. Biotin, or Vitamin H, is also known as a growth-promoting

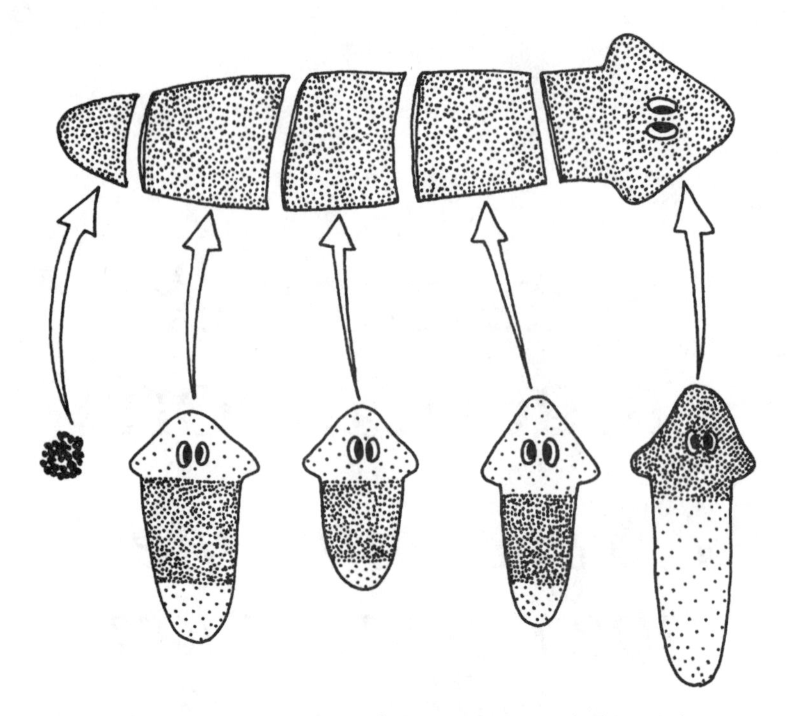

17-1 *Planaria possess one of the most remarkable abilities to regenerate lost body parts of all the animals.*

vitamin. It is commonly found in egg yolks and beef liver, two foods that planarians eat.

This project tries to determine if Vitamin H (biotin) is beneficial to a planaria's ability to regenerate lost body parts and, if so, which concentration is the most beneficial. Biotin is considered a growth-promoting vitamin in humans. Is it important in regeneration capabilities of lower forms of life such as the planaria?

Materials list

- Biotin as powdered capsules, 1000 mcg each (available at pharmacies)
- Cardboard box with cover (dark place for petri dishes containing planarian)
- Sterile, mini-brushes (for transferring the planaria)
- One 4 oz. glass bottle
- One 10 oz. glass bottle
- Graduated cylinder
- Ice cubes
- Spring water

- Forty white, self-sticking labels
- Magnifying glass
- Dissecting scope (optional)
- Medicine dropper
- Three to five microscope slides
- About 50 petri dishes
- About 40 planarians (Black planarians were used in the original experiment, but others should work fine.)
- Food for the planarians (depends on the species of planaria used)
- Five to ten sterile razor blades

Procedures

First, you will make four experimental solutions containing vitamins to be used over the course of the project, which will last for one week. These solutions will be used to change the planarian's water four times during that week, using 30 milliliters of solution each time.

Create the following solutions, and label each bottle with the solution letter. Solution A is the control. It contains 120 milliliters of pure spring water and no biotin. Solution B contains 120 milliliters of spring water plus the contents of three 1000 mcg biotin capsules. Mix thoroughly. (See Fig. 17-2.)

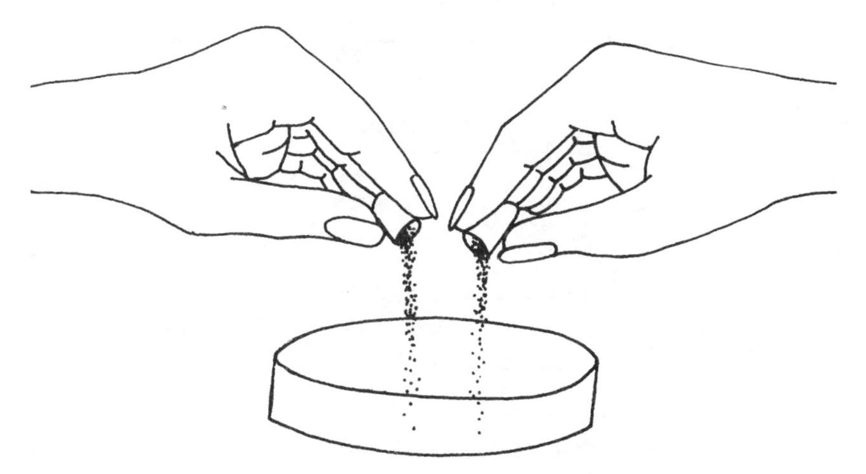

17-2 *Pour the contents of each vitamin capsule into the petri dish.*

Solution C contains 120 milliliters of spring water plus the contents of four 1000 mcg biotin capsules. Mix thoroughly. Solution D contains 120 milliliters of spring water plus the contents of five 1000 mcg biotin capsules. Mix thoroughly.

Label each of four petri dishes with "A," "B," "C," or "D." Add 30 milliliters of each solution prepared above in the proper dish. Then transfer nine planarians into each dish. Keep the planarians in the solutions for one week, cleaning and refilling the petri dishes every other day with the proper solution.

Feed all the planarians the same amount of food each day. The type of food depends on the type of the planarian. Brown planarians like hard-boiled egg yolks, and black or white planarians like fresh liver. Tubifex from a pet store can be used to feed most planarians. Always remove the food 30 minutes after feeding. The best time to refresh the solution is after the food has been removed.

On the eighth day, you will prepare the planaria for the regeneration study. You will make four different types of cuts for each of the four solutions. Two sets of these four cuts will be made for each solution. (There will be eight planarians used in each solution.) The four cuts are: half, straight down the middle, between the eyes, and three ways. (See Fig. 17-3.)

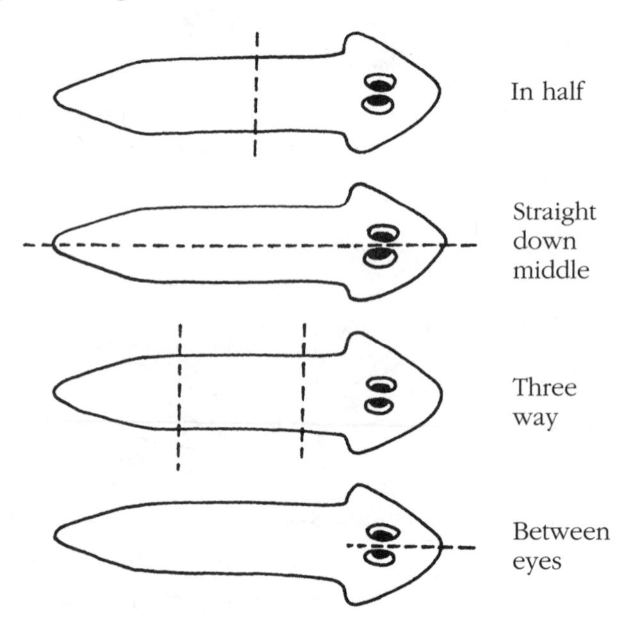

17-3 *Cut the planaria in the following four ways.*

To cut the planarians, place a planarian on a microscope slide with a drop of spring water on it. Place the microscope slide on an ice cube. This will cause the planarian to extend its body making it easier to cut. A dissecting scope or someone holding a magnifying glass for you will make it easier to see the subject while you make the incision.

Place each piece of planarian in its own petri dish with 30 milliliters of pure spring water. Label the dish with the solution it had been living in for the past week (A, B, C, or D) and the cut it had received. Don't feed the planarian during this period, but change the water every other day. Observe all the planarian pieces each day. Note their overall health, color, and degree of activity and the progress of regeneration. Note the date and time of any deaths. Follow the progress of all the planarians for six weeks.

Analysis

How did the experimental specimen's regeneration compare with the control, which received no biotin preceding the cut? Does it appear that access to biotin prior to the cutting helped the planarians regenerate? If so, which concentration appeared to be the most beneficial? Were any concentrations harmful?

Going further

Continue this project to see if biotin would be beneficial if administered during the actual regeneration process instead of prior to it. You can also test other vitamins and/or minerals. Are they beneficial or are some harmful, if not toxic, to planarians? You can also run similar tests on other organisms capable of regeneration, such as the sea star or starfish.

Suggested research

- Research the biochemical activity of biotin. How might it be involved in regeneration?
- Do a comparative literature search on regenerative abilities of animals. How and why do these powers differ so between lower and higher forms of animal life?

Part 5

Animals & our environment

Over the past few decades, we have polluted and degraded our environment to the point that many scientists are concerned about the future of our planet. When one organism is affected by pollution, a ripple effect is usually felt throughout the local ecosystem and often beyond into more distantly interconnected ecosystems.

What happens to one organism is often an indication of things to come for many others. This part contains projects that investigate the effects of pollutants on animals and looks for options.

The first project investigates the most common indoor pollutant, second-hand smoke. What is the effect of second-hand smoke on a terrestrial organism, such as a mealworm?

The next project looks into whether a new, safer type of herbicide, called a growth regulator, might not be as safe as many think.

Automobile batteries contain lead, household batteries contain mercury or cadmium, and manufacturing processes use numerous other heavy metals. These heavy metals are toxic to life. When they are disposed of, they infiltrate the land, the water, and the air. This third project investigates the effects of a heavy metal on an aquatic organism, brine shrimp.

Next, we will return to second-hand smoke and study its effects on very different types of organisms: mosquitoes and man.

When fossil fuels are burned, they release fly ash, which contains many pollutants. The next project investigates the effect of coal fly ash on mealworms.

The final project in this part—and in the book—investigates population dynamics by following the growth of a population and theorizing about its potential.

18

Second-hand smoke

Does second-hand smoke affect mealworms?

The inhalation of cigarette smoke by nonsmokers is called *second-hand smoke.* (It is also called environmental tobacco smoke, ETS, and passive smoke.) The Environmental Protection Agency and the Surgeon General's office have stated that this smoke causes illness and probably deaths. A recent study, which refers to over 50 other studies, concludes that second-hand smoke causes 3,000 lung cancer deaths per year.

Project overview

This project attempts to show the hazards of second-hand smoke by demonstrating the threat it poses to the health of organisms other than humans. This project uses mealworms, but a similar experiment can be modified to accommodate virtually any type of organism.

The purpose of this research is to analyze and assess the effect of second-hand cigarette smoke on mealworms. The mealworms are subjected to second-hand smoke for a period of time. Then, their degree of health is measured by noting their length, weight, and mortality rates.

Materials list

- Sixteen standard petri dishes
- Approximately 100 mealworms, *Tenebrio molitar*, in the larvae stage (available at pet stores)
- Two 15-cm pieces of 4-mm plastic tubing (standard air line tubing available at pet stores)
- Two large serving platters with large plastic bubble type tops (the kind used to serve a turkey)
- Two jars with plastic screw cap lids (such as peanut butter jars)
- Package of cigarettes (These must be obtained and monitored by your sponsor or parent throughout this project.)
- Twenty grams of dry bran (such as Quaker Oats)
- Standard cm ruler to measure the mealworms' length
- Scale or balance to measure the mealworms' weight
- Caulking material for plastic (available at hardware stores)
- Drill with a bit wide enough to make holes five mm in diameter (to drill through thick plastic)
- Electrical outlet
- Aquarium air pump (available at pet stores)
- Five-foot extension cord
- Book of matches
- Twelve-by-twelve-inch piece of foam (available at sewing shops)

Procedures

First, you must prepare the mealworms. Remove the covers from all 20 petri dishes. Use the scale or balance to measure one gram of bran, and place that amount in each dish. Then, place five mealworms in each dish. Label 12 of the dishes "experimental," and four of them "control." Place all the covers back on the petri dishes. Place the petri dishes that are labelled "control" in an environment where there is no second-hand smoke. This room will also be used for the experimental groups after they are removed from the smoke chamber.

Preparation of the second-hand smoke apparatus in which the "experimental" petri dishes will be placed is as follows. Drill one hole, 5 mm in diameter, in the top of the serving platter plastic top. Drill two 5-mm holes in the plastic lid of the jar. (See Fig. 18-1.)

Insert one end of one of the pieces of tubing into one of the holes in the jar cover and the other end into the hole in the serving platter. Use a caulking gun to create a seal where the tubing enters the holes.

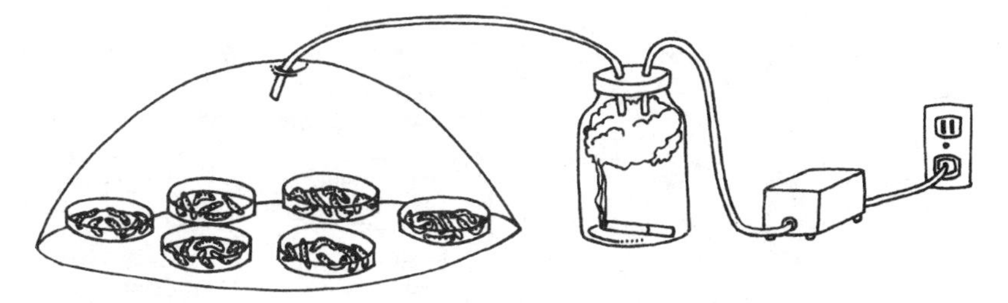

18-1 *You will create a second-hand smoke chamber as shown here.*

Use the second piece of air tubing to attach the air pump to the other hole in the lid of the jar as seen in the figure. Use the caulking gun to create a seal. Let the caulk dry overnight.

The air pump will pump air through the cigarette chamber, carrying with it the smoke from the cigarette. The air will continue through the tubing and enter the serving platter where the petri dishes containing the mealworms will be placed. The smoke will accumulate under the serving platter.

When the apparatus is ready, plug the air pump into an electrical outlet and turn on the pump. Have an adult light the cigarette and place the lit cigarette in the jar as shown in the figure. Attach the jar to the cover (which has two air lines attached).

Take four of the petri dishes labelled "experimental," and add to the label "24 hours." Place these four petri dishes into the serving platter and cover with the top (which has an air line attached).

This first group will be exposed to the second-hand smoke for 24 hours. (During the experiment, you will have to take measurements as described below.) An adult will have to relight new cigarettes every two hours to maintain the smoke. (The smoke will remain in the chamber for a couple of hours after the cigarette has become extinguished.)

After 24 hours, remove the petri dishes. Prepare the second group of four petri dishes by adding "12 hours" to the labels. (See Fig. 18-2.) Place these dishes in the chamber and begin the experiment, but run it for only 12 hours. Repeat this procedure for the next group of four petri dishes, but run it for only 6 hours.

To take measurements, use a cm ruler to measure every mealworm at the beginning of the study, every sixth hour into the test, and at the conclusion of the test. To measure the length of each mealworm place the ruler down on a flat surface. Lay the mealworm out on the ruler, straighten it, and note the length. To measure the weight, use a scale to measure the weight of each group of five mealworms and divide by five. Be sure to take into consideration the weight of the petri dish when using the scale or balance.

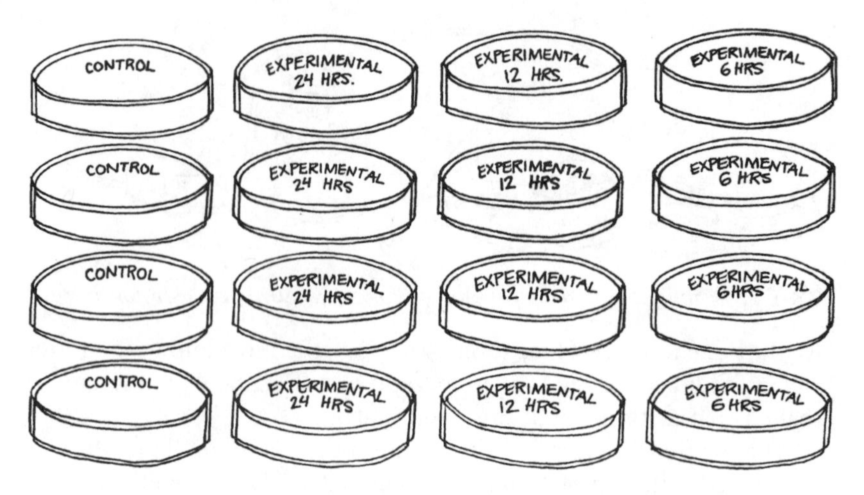

18-2 *You'll have four sets of four petri dishes with the following labels.*

To measure mortality, count the amount of mealworms that die by using the poke test every sixth hour. To perform the poke test, turn the mealworm belly-side up and use your finger to gently touch the middle of the abdomen. (See Fig. 18-3.) If it doesn't respond with a jerking motion, assume it is dead.

You should also record the general condition of the living mealworms. Note their color, movement or lack of it, and any other signs that might indicate their health (or lack of it).

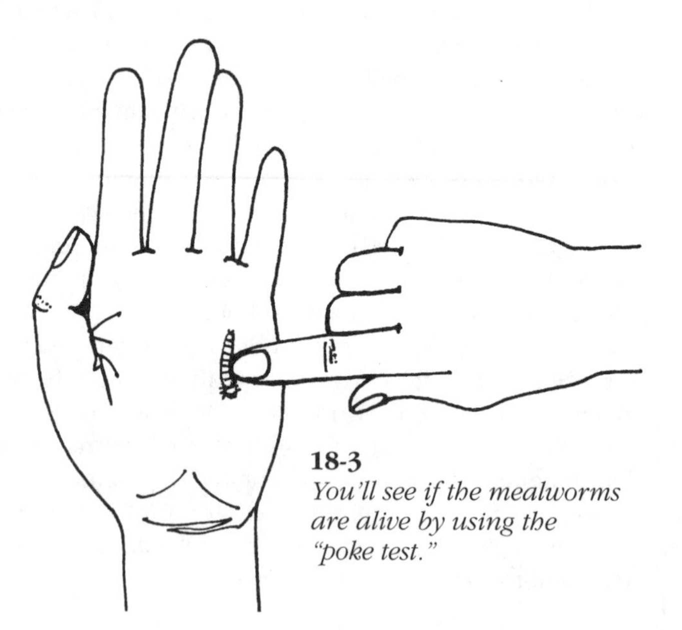

18-3 *You'll see if the mealworms are alive by using the "poke test."*

Analysis

Did any of the measurements indicate a decline in the health of the mealworms for any of the groups? Does it appear that the passive smoke had a negative effect on the insect larvae? Was there any difference between the experimental groups? Did it take a lot of second-hand smoke to get results or does it take very little smoke to cause ill effects?

Going further

This project only tested the short-term effects on the mealworms. Continue this project to see the long-term effects of second-hand smoke. For example, allow the mealworms to pupate and emerge as adults. Then examine the condition of the adults. Are they all the same size? Did the same percentage from each group emerge as adults?

Suggested research

- Research the latest studies on second-hand smoke. There has been an explosion of information about this problem in the past year.
- Research how smoke has been used on occasion to control insect pests.

19

Growth regulator herbicides

Do these new herbicides affect planaria regeneration?

Herbicides are pesticides that specifically kill unwanted vegetation. When used by the homeowner, they are called weed killers. They are, however, primarily used in agriculture to keep unwanted vegetation from interfering with crops. They are also used on a large scale to clear vegetation along road sides, under high-power tension wires, and any other place where vegetation won't be tolerated. By far, most pesticides applied are herbicides. About 65% of all pesticides applied in the U.S. are herbicides.

Herbicides, like all pesticides, are toxic to many forms of life other than the intended, or targeted, weed. Most of them kill a wide range of plants and harm many types of animals, especially aquatic organisms. A relatively new type of herbicide that is capable of destroying only the targeted weed has been developed. These herbicides use the plant's own growth hormones to control them. Synthetic hormones mimic the actual plant hormones. When applied, they interrupt the plant's natural growth. For example, when some growth hormone herbicides are applied to pest plants, it makes the plants grow so rapidly they fall over on themselves and die. Others make the plant stop growing.

The advantage these growth hormone herbicides have over conventional chemical herbicides is that they are more specific, usually only attacking the targeted pest plant. But are these new herbicides harmful to animals that might be found living alongside the weed being destroyed?

Project overview

This project will test to see if gibberellic acid, a commonly used growth hormone herbicide, has any effect on the regeneration abilities of *Planaria*, the common flatworm, found living in freshwater streams. If it has a negative effect, should we not be concerned about the overall effect these new types of herbicides have on the environment?

Materials list

- Gibberellic acid (available from a scientific supply house)
- Forty *Planaria* (They can be caught or purchased from a scientific supply house.)
- Petri dishes
- Magnifying glass
- Stereoscope
- Stream
- Distilled water
- Eight test tubes
- X-acto knife
- Two quart-sized jars with screw cap lids
- Tap water
- Pipette

Procedures

If you purchased your planarians, skip to the next paragraph. If you want to catch your own planarians, follow these instructions. Go to a nearby freshwater stream with two clean jars. Fill the two jars half full with stream water and set them aside. Pick up a small rock that has been submerged in water and shake the rock in the water in one of the jars. Look on the rocks for planarians. A magnifying glass will make it easier to see them. Tweezers might help get them into the jars. The other jar will be used to bring back spring water.

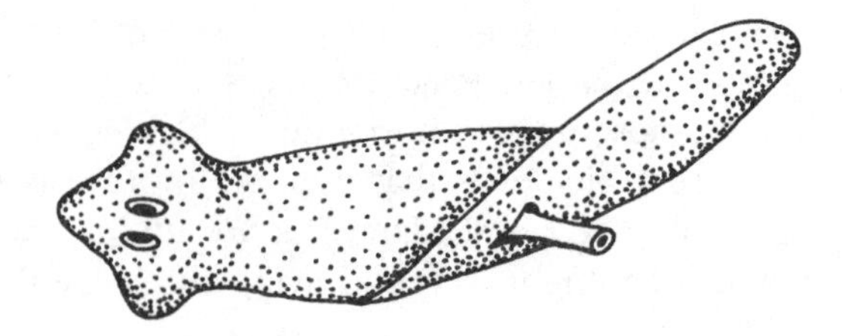

After you have obtained your planarians, bring them to the lab and transfer them into a dish containing spring water using a pipette. Using the stereoscope, separate the planarians from any other freshwater animals you might have collected. Next, prepare the petri dishes and test tubes for use. Label the test tubes from #1 to #8. Put ten ml of distilled water into test tube #1. This will be your control. Label it as such.

Then, put ten ml of undiluted gibberellic acid into test tube #2. In test tube #3, take one ml of the solution from test tube #2 and put it into test tube #3, and add nine ml of water. This will create a 1:10 dilution. Label the test tube as such.

In test tube #4, take one ml of the solution from test tube #3 and add nine ml of water. This makes a 1:100 dilution. Label this test tube as such. Repeat these procedures for the remaining test tubes. You'll have a 1:1000 dilution in test tube #4 and a 1:1000000 dilution in test tube #8. (See Fig. 19-1.) Label all the test tubes.

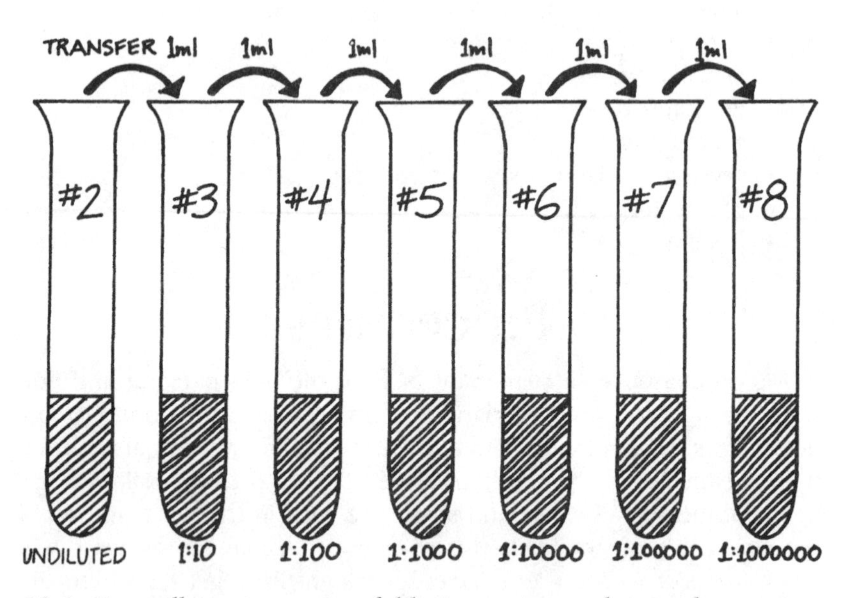

19-1 *You will create a series of dilutions using each test tube to create the next dilution.*

Now that your test tube solutions are ready, you'll prepare the planarians for the regeneration test. The best way to cut the planarians is to place them on a microscope slide and place the slide on a piece of ice under the stereoscope. Because of the cold, they will extend themselves, making it easier to cut them. Using an X-acto blade, cut each planarian in half, horizontally. Using the pipette, place ten halves into each of the eight numbered test tubes. The planarians are not fed during the regeneration process.

Begin collecting data on the third day after sectioning the planarians. Your data should show how many planarians are alive, how many planarians showed signs of regeneration, and if any have died and are decomposing. At the conclusion of 18 days, which is six observations, gather and compile the data.

Analysis

Did any of the dilutions of gibberellic acid interfere with planarian regeneration or result in more deaths? Compare the dilutions with the control. Graph the number of deaths and the speed of regeneration for each dilution over the 18 days. Did higher or lower concentrations result in slower regeneration or more deaths over time? Does gibberellic acid only interfere with the growth of its targeted pest plant or does it have a greater effect on the ecosystem?

Going further

Gibberellic acid is just one of many types of growth-regulator hormone herbicides. Scientific supply houses sell kits containing about a dozen of these types of hormonal herbicides. Continue the research by using a variety of these herbicidal hormones or use a different experimental organism such as brine shrimp, nematodes, or earthworms.

Planaria is not the only organism capable of extensive regeneration. If you have access to a marine (saltwater) aquarium, try to run a similar test on starfish or sea stars. (See Fig. 19-2.) Are the results similar in both fresh and marine environments?

Suggested research

- Research the scientific literature about the environmental impact of growth regulator herbicides, especially on aquatic ecosystems, and compare it with the environmental impact of synthetic chemical herbicides.
- Look into the use, misuse, or lack of use of growth regulator herbicides by farmers.

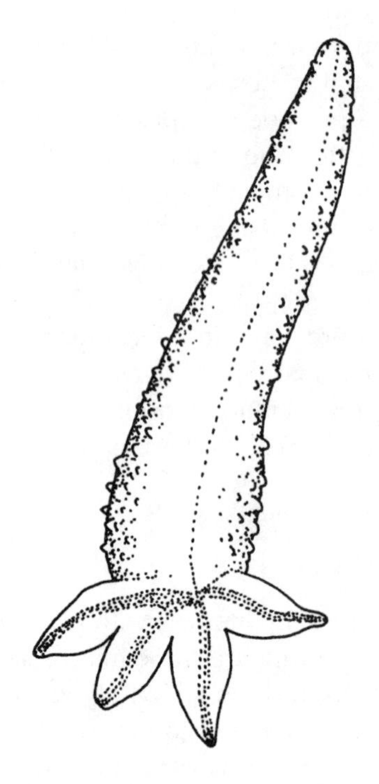

19-2
The planaria is not the only organism capable of amazing feats of regeneration. Here a detached sea star arm is regenerating an entire new body.

20

Heavy metal pollution

Does copper sulfate pollution harm brine shrimp?

Industrialized nations of the world manufacture vast quantities of materials that contain substances harmful to humans and the environment in general. These substances are considered toxic if they pose a health threat to any organism including people. One group of toxic substances that many scientists and environmentalists are becoming increasingly concerned about are heavy metals.

These metals include lead found in car batteries and some paints, mercury found in some batteries and in marine paints, and cadmium found in rechargeable batteries. Most of these heavy metals cause disease and death if found in high enough concentrations.

Project overview

Copper is found in some paints, pipes, and other industrial products. Copper, like most heavy metals, is a toxic substance. When products containing copper are disposed of or disintegrate, the copper enters into the environment. It can end up in a pond, in the soil, or in our drinking water. What is the effect of copper solutions on an aquatic organism such as brine shrimp? If copper can harm brine shrimp, which are found at the lower end of many aquatic food chains, it would have a serious impact on the entire aquatic ecosystem. In addition, if copper in very low concentrations is harmful to brine shrimp, what might happen if humans ingest this metal in drinking water?

In this project you will test varying concentrations of copper sulfate in water to see if there is a negative effect on brine shrimp.

Materials list

- About 75 liters of spring water with no salt added (available at supermarkets)
- About 4 kg of non-iodized salt (available at supermarkets)
- Copper sulfate (available from a scientific supply house)
- Thirty-five large petri dishes (8.5 cm diameter)
- Balance
- Electronic balance accurate to ¹⁄₁₀₀₀ g (might be available at your school)
- Graduated cylinder (10 ml)
- Graduated cylinder (1000 ml)
- Funnel
- Marking pen
- Plastic trash bags
- Latex gloves
- Parafilm (might be available at your school)
- Four large beakers
- Fish bowl or aquarium to maintain the shrimp
- Live adult brine shrimp (available at pet stores)
- Brine shrimp food (*Artemia*)
- About 40 cm of small plastic tubing
- Air pump
- Air stone

Procedures

You will first prepare the copper sulfate solutions. Add 3,000 ml of spring water into each of the four beakers and set aside. Using an electronic balance, measure out the following three quantities of copper sulfate: 0.9 g, 0.09 g, and 0.009 g. Add the 0.9 g of copper sulfate to the first beaker, the 0.09 g to the second beaker, and the 0.009 g to the third beaker. Since the fourth beaker will be the control, add nothing to it.

WATER	+	$CuSO_4$	+	SALT	+	($CuSO_4$) mg./liter (WATER)
3,000 gr.		.9 gr.		90 gr.	=	"300"
3,000 gr.		.09 gr.		90 gr.	=	"30"
3,000 gr.		.009 gr.		90 gr.	=	"3"
3,000 gr.		0 gr.		90 gr.	=	"CONTROL"

The beaker with the 0.9 g of copper sulfate added now has 300 mg of copper sulfate per liter of water, so label it "300." The one that has 0.09 g has 30 mg per liter, so it should be labelled "30," and the one with 0.009 g has 3 mg per liter, so label it "3." Label the control beaker, "control."

To prepare these four beakers for the brine shrimp, you must adjust each beaker to a 3% salt solution. Do this by weighing out 90 grams of non-iodized salt on the balance or scale and add that amount to each beaker. All four beakers are now slightly saline and contain varying concentrations of copper sulfate, except for the control which has no copper sulfate.

Now, you will prepare the brine shrimp for the experiment. Put 3,000 ml of spring water and 90 grams of non-iodized salt into the empty fish tank. The tank now contains a 3% saline solution. Allow the fish tank to reach room temperature. Attach the tubing to the air pump and an air stone to the other end of the tubing. Plug in the pump, and insert the air stone into the tank to aerate the water. You can now add the live adult brine shrimp. (If you are rearing your brine shrimp, follow the instructions that came with the package, or ask your sponsor for instructions.)

Feed the brine shrimp once each day with two drops of food. If you are using *Artemia*, which is the common food sold in pet stores, you will need to add one scoop of *Artemia* with 200 ml of water and mix thoroughly.

You are now ready to begin the actual experiment. Use a marker to label five petri dishes for each beaker prepared earlier. You'll have five petri dishes labelled "control," five labelled "300," and so on. Measure out 20 ml of solution from each of the beakers, and place it in the properly marked petri dishes. For example, measure out 20 ml from the "30" beaker and place it in one of the petri dishes labelled "30." (See Fig. 20-1.)

After all the petri dishes contain the proper solutions, add three brine shrimp to each dish. (You'll have a total of 20 petri dishes, so

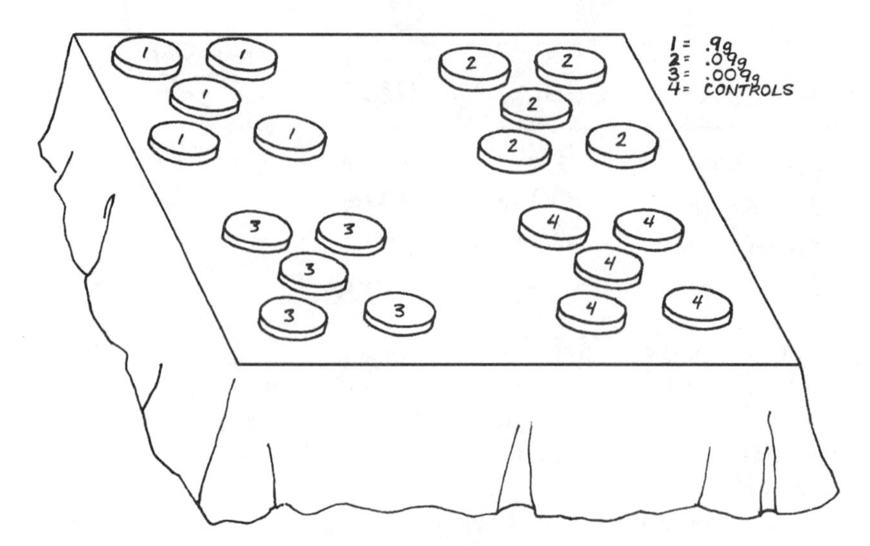

20-1 *The four solutions will be set up as shown.*

you'll need 60 adult brine shrimp.) Begin recording the number of shrimp that are alive and those dead in each dish every day for one week. Continue feeding the brine shrimp as usual.

Analysis

Plot a graph to analyze the data. (See Fig. 20-2.) What was the effect of copper sulfate on the adult brine shrimp? If there was a negative effect, did it occur at all concentrations or just the higher concentrations? How did the three experimental groups compare with the control group? Did those in any of the experimental groups do better than those in the control. If there was a negative effect, did it occur immediately or after a few days? Does it appear that all concentrations of copper sulfate are harmful to life or just some?

Going further

Run a similar experiment using young, immature brine shrimp. Use lower concentrations. What would happen if young brine shrimp are more severely affected than adults? Would it cause more or less harm to the species and the local ecosystem?

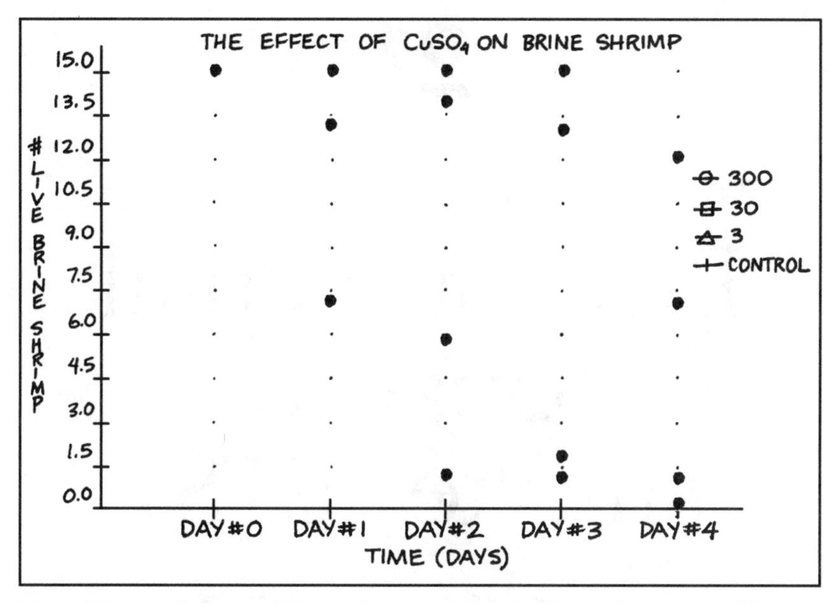

20-2 *Plot a chart similar to this to see the effect of copper sulfate on brine shrimp.*

Suggested research

- Research the sources of heavy metal pollution in our environment and the harm it causes.
- Find information about how the government and businesses are trying to reduce the use of many heavy metals and eliminate the most harmful ones.

21

Environmental tobacco smoke (ETS)

Is it harmful to all forms of life?

It is called passive smoke, environmental tobacco smoke and second-hand smoke. All of these terms refer to smoke inhaled not by the smoker, but by those around him or her. The negative health effects of smoking have long been known. The dangers of passive smoke, however, have only recently been investigated.

A massive amount of research has begun on this topic. The Surgeon General's Office believes that tens of thousands of deaths can be attributed to passive smoke.

Project overview

In this project, you will survey the harmful effects of passive smoke on two distantly related animals. An insect has a simple respiratory system consisting of a series of tubes, called tracheae, that carry gases to and from all the cells in the body. (Blood has nothing to do with the delivery of gases into an insect's body.) Tracheae reach the outside of an insect's body through openings called spiracles. (See Fig. 21-1.)

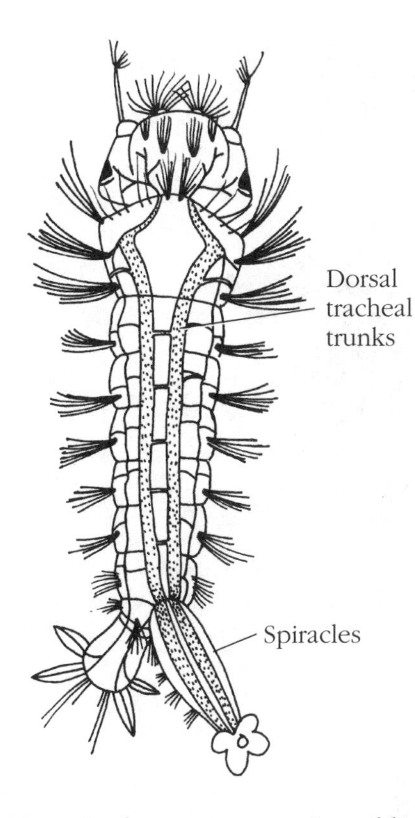

21-1
Mosquito larvae use special spiracles and a tracheal system to deliver gases throughout their bodies instead of the circulatory system.

Mosquito larvae are aquatic and live just beneath the water's surface. They have special breathing tubes that pass upwards, through the water's surface, allowing air to enter. (See Project 15 for an experiment about passive smoke on terrestrial insects.)

At the other end of the spectrum, we have humans with more complex respiratory systems. In the human respiratory system, air is delivered to the lungs, which then transfer gases to the circulatory system where the blood completes the trip. (See Fig. 21-2.)

When people exercise, their heart rates increase and, in turn, their pulse rates increase. When a person stops exercising, the heart gradually slows back to normal speed and the pulse returns to normal. The amount of time it takes for the pulse to return to normal is called the recovery rate. The recovery rate is a key factor in determining a person's health.

How does passive smoke affect these two forms of life? You'll see how passive smoke affects the mortality of mosquito larvae and how it affects the pulse recovery rate in people. Is there a significant difference in the recovery rate between those who regularly inhale passive smoke and non-smokers?

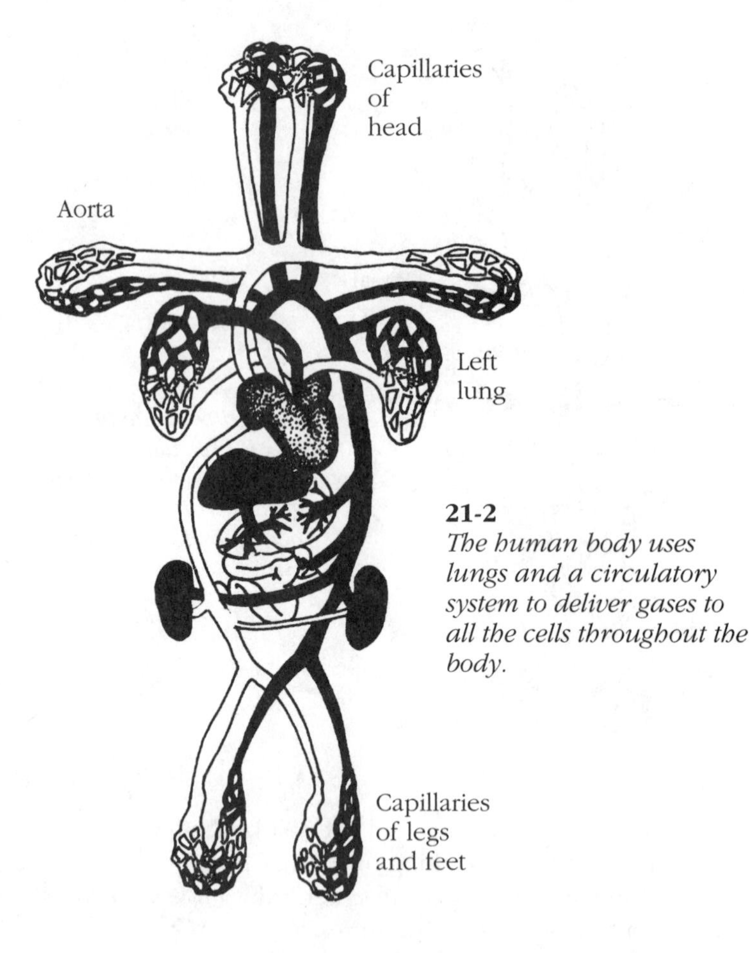

21-2
The human body uses lungs and a circulatory system to deliver gases to all the cells throughout the body.

Materials list

For passive smoke and humans:

- A group of at least eight non-smoking adults who do not work or live with people who are smokers, all about the same age
- A group of the same number of adults who live or work with regular smokers, all about the same age (all individuals should be in good health and have permission from the doctors to participate)
- Stopwatch or watch with a second hand

For passive smoke and mosquito larvae:

- Three and a half feet of plastic tubing (½ inch diameter)
- Two plastic or rubber food containers (about quart size) with air tight, sealed covers (You'll have to put holes in the covers.)
- Hot glue gun to seal the holes
- About 50 mosquito larvae (You can collect these in the summer by placing a bowl of water containing a few dead

leaves outside for a few days, or they can be ordered from a scientific supply house; if you order them, get either *Aedes* or *Anopheles*, not *Culex*.)
- Mosquito food (This can be ordered with the mosquitoes, or you can use flake goldfish food that you've crushed into a dust.)
- A few cigarettes with filters (the use of cigarettes in this project must be monitored by your sponsor or parent.)
- Matches
- Cotton
- Two small pieces of mesh (such as screening)
- Small rubber bands or packing tape
- Magnifying glass
- Utility knife
- Marker
- Small piece of aluminum foil

Procedures

Part one

For the first portion of this project follow these procedures. Have the first individual from the non-smoking group sit and rest for three minutes. Then take his or her pulse for one minute and record the results. This is the "at rest" reading. Then have this person jog in place for two minutes. As soon as he or she finishes jogging, take his or her pulse once again and start the stop watch.

Wait two minutes and take the person's pulse once again. Record the pulse at the two minute mark. Take the pulse again every other minute, recording the pulse and the time elapsed. Do this until the pulse rate has returned to normal. (The "at rest" rate.) Repeat this procedure for the other non-smokers. Then repeat the procedure for the passive smokers. Once all the data has been collected, average all the data for each group.

Part two

For the mosquito larvae portion of this experiment, you must first create the containers to hold the passive smoke and the developing mosquitoes. Use the utility knife to cut two holes just big enough to fit the plastic air tubing through into the lids of each bowl. Cut the plastic tubing into two 12-inch lengths and two 6-inch lengths. Insert a 6-inch length into one of the holes you just cut and pull it through so it almost reaches the bottom of the bowl. There should be only an inch or two sticking out of the bowl.

Next, insert a 12-inch length into the other hole. It should extend into the bowl only about one inch with the remainder sticking out of the bowl. Once the tubes are properly positioned, use a hot glue gun to form an airtight seal around the tubes and the top. Repeat this procedure for the second bowl. Mark one of the bowls "smoke" and the other "control." (See Fig. 21-3.)

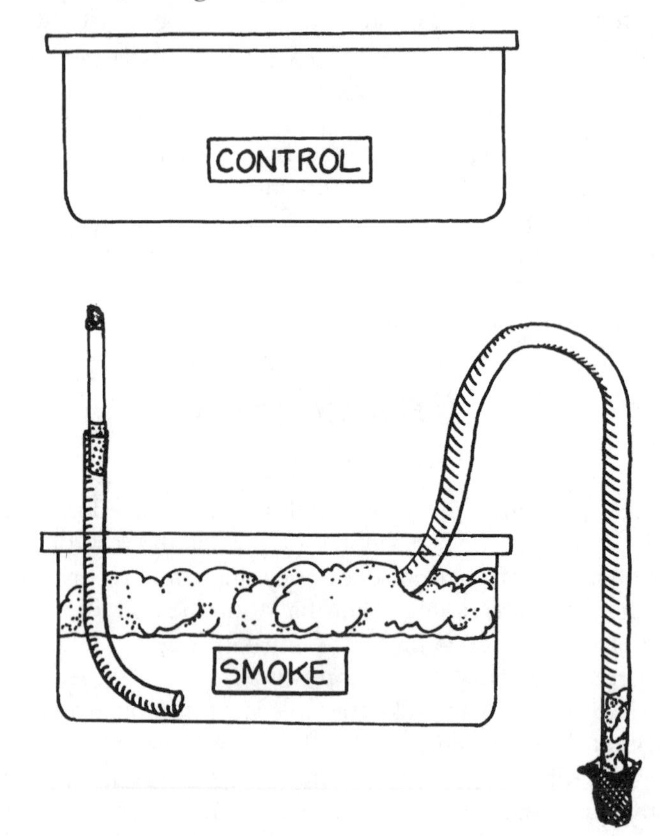

21-3 *The smoke chamber looks like this.*

Put about 1.5 inches of pond water into each bowl. If pond water isn't available, let tap water sit for at least 24 hours before using it. Be sure the 6-inch tube reaches well into the water and the 12-inch tube remains well above the water as you see in the illustration. Place 25 young mosquito larvae into each bowl, add a pinch of food and cover both bowls.

Place small wads of cotton into the ends of the 12-inch tubes. Make it tight enough so that it can't be easily forced out. Cover the tube opening (containing the cotton plug) with a small piece of mesh and secure it with a rubber band or tape wrapped around the tube. (This is the tube you'll be drawing air through so you want to make sure the cotton isn't sucked into your mouth.) Most of the smoke will be caught by the cotton, so it won't enter your mouth, and will be trapped in the bowl. (It's inserted into both bowls for consistency.) Insert the filter end of a cigarette into the end of the shorter tube on the bowl labelled "smoke." Leave this tube open in the "control" bowl.

You are now ready to begin the experimental phase of the project. Place a match to the cigarette. Inhale through the long tube that contains the cotton plug. As you inhale, the cigarette will light and smoke will be carried through the small tube and bubble through the water. Some of the smoke will continue to pass through the long tube and be stopped by the cotton plug. (Almost all the smoke will be stopped by the cotton plug, but do not inhale any smoke that might pass through the cotton.) Once you are sure the apparatus is working properly, inhale through the tube ten times. When done, remove the lit cigarette from the tubing and dispose of it safely. The air in the bowl should be filled with smoke.

Now, repeat the inhaling procedure with the "control" bowl, without any cigarette. Only air will pass through the water. You'll still see bubbles, but there won't be any smoke. Inhale ten times, just as you did with the "smoke" bowl. Take notes. Leave the bowls undisturbed for 24 hours and then repeat this procedure. You'll continue these steps for seven days. Feed the larvae once or twice each day. Use the magnifying lens to observe the larvae.

Look for dead larvae in both bowls. Look for larvae that have pupated. Once you find pupae, continue the experiment to see how many emerge as adults. Once adults have emerged, release them outside.

Analysis

In the first part of the project, what was the pulse recovery time for the non-smoking and the passive smoking group? Was there a significant difference between the two groups? Does passive smoking appear to affect the ability of the heart to recover from exercise as indicated by the recovery pulse rate?

For the second part of the project, how did the passive smoke affect the mosquitoes? Was there more mortality in the "smoke" bowl than in the control bowl? Did they show any ill effects? Did the same number of mosquito larvae pupate in both bowls?

Going further

For the first part of the project, compare differences in pulse recovery rate between passive smokers and actual smokers. Is there a greater difference between the non-smoker and passive smoker or passive smoker and actual smoker?

For the second part, devise an experiment that would study whether passive smoke affects adult mosquitoes after they emerged. Look at size and weight differences between adults that "smoked" while larvae and those that didn't.

Suggested research

- What does the latest research on the dangers of passive smoking reveal?
- Does cigar or pipe smoke differ in its danger as passive smoke?
- Can smoke be used to control mosquitoes in the wild? Is it being used today?

22

Fossil fuels & animals

Does coal fly ash harm mealworms?

Although alternatives do exist, most of the electricity generated in the U.S. and the world comes from burning fossil fuels, such as coal, oil, and natural gas. When fossil fuels are burned, they cause air pollution, including fly ash and ground level ozone, commonly called smog.

Fly ash is the "soot-like" substance that goes up the smoke stack and blackens the sky if proper anti-pollution devices, such as electrostatic precipators, are not in operation. Fly ash is also produced when municipal solid waste, or garbage, is incinerated. Burning fossil fuels or garbage reduces the volume of what is being burned by 60% to 90%. But what happens to the fuel or the garbage being burned? Where does it actually go?

Most of it goes up the smoke stack to pollute the air. This is called fly ash. The worst fly ash is produced by low-grade, coal-burning power plants. Coal is the most common type of fuel used to generate electricity. This coal fly ash is carried by wind currents and deposited across the land and the sea where it can affect people's health and the environment in general.

Project overview

Coal fly ash contains small particles that interfere with the normal functioning of many kinds of animals, from people, in which it clogs the lungs, to small insects, in which it clogs the tracheal system.

This project investigates the effect of coal fly ash on mealworms. These common terrestrial insect larvae can be used as an indicator species to see if varying levels of coal fly ash can interfere with an organism's normal growth and development.

Materials list

- Twelve petri dishes
- Teaspoon of oatmeal
- Four measuring spoons (⅛, ¼, ½, 1 teaspoons)
- Six paper towels, cut in circles
- Forty-eight mealworms (available at pet stores)
- Three ¾ teaspoons of coal fly ash. (Get permission to obtain the ash from the manager of a electric power plant. If it's not available, you can use ash from a charcoal grill.)
- Metric ruler
- Two pencils and a rubber band to measure the worms

Procedures

You'll prepare two sets of six groups of petri dishes as follows. Label the first two petri dishes "Group One." Put ⅛ of a teaspoon of coal fly ash, ½ teaspoon of oatmeal, and ⅛ teaspoon of water into each of the two dishes and mix. Label the next two, "Group Two" and mix ¼ teaspoon of coal fly ash, ½ teaspoon of oatmeal, and ⅛ teaspoon of water in each. For the next two, label them "Group Three," and mix ¼ teaspoon of coal fly ash, ½ teaspoon of oatmeal, and ⅛ teaspoon of water in each. "Group 4" will get 1 teaspoon of coal fly ash, ½ teaspoon of oatmeal, and ⅛ teaspoon of water mixed in each.

Label the next two dishes "Control 1," and add 2 teaspoons of oatmeal and ⅛ teaspoon of water to each. The final two dishes are labelled "Control 2," they get only ½ a teaspoon of oatmeal each. Place four mealworms in each petri dish. All the dishes are kept at room temperature.

Measure the length of the mealworms at the beginning of the project and every five days thereafter. To measure the length of the mealworms, place two pencils together and wrap a rubber band around them. (See Fig. 22-1.)

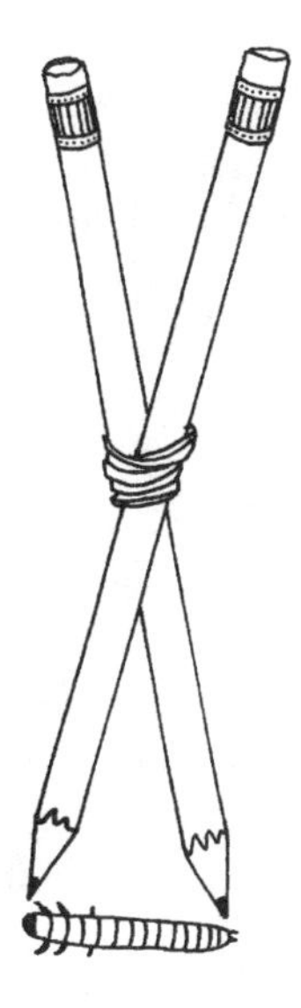

22-1
Wrap a rubber band around the middle of two pencils and hold the points at the ends of the mealworm.

Take one mealworm out of its petri dish, and place it on a paper towel. Adjust the pencils to the actual length of the insect as shown in the figure. While maintaining the distance between the pencil points, place the pencils points on the ruler and record the distance between the two pencil points. (See Fig. 22-2.)

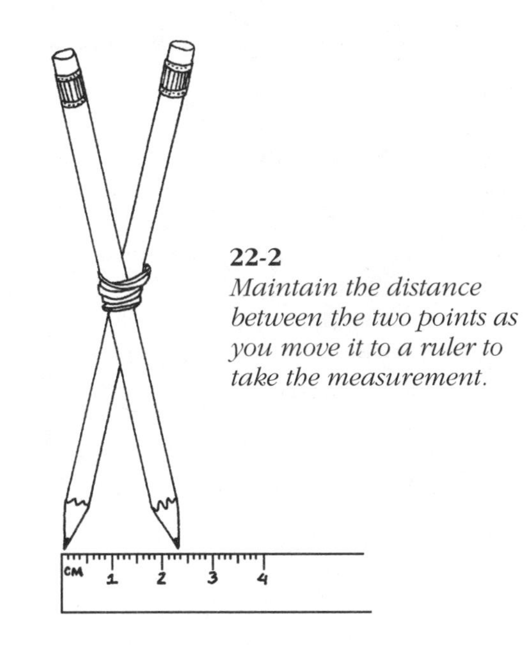

22-2
Maintain the distance between the two points as you move it to a ruler to take the measurement.

Record the death of any mealworm. Also note the overall condition of the mealworms by observing their color, movement, and so on. Continue your observations until the mealworms pupate. Record the date when pupation occurs for each mealworm, and then calculate the average for each dish and each group (of two dishes). Take a final reading by measuring the length of the pupal case.

Analysis

What percentage of the total number of larvae in each group died? Did a greater percentage die in the experimental groups than in the control group? How long did it take for the larvae to pupate? Did the experimental groups take longer than the control groups? Was there a difference in the average length of individuals in different groups? Does it appear as if coal fly ash is detrimental to mealworms in any amount or only at high levels?

Going further

Allow the pupae to emerge as adults, and see if there is a difference in the adults' size and activity. The adults can be dissected to see if there are any anatomical differences between those exposed to fly ash and those that were not exposed.

Suggested research

- Investigate the effects of coal fly ash on other organisms, including people.
- Research how coal fly ash and incinerator ash are being controlled.
- How, and to what degree, do pollution control devices reduce the amount of ash leaving the smoke stack?
- Determine actual amounts of ash that might affect the lives of organisms. For example, how much fly ash might an insect living near a coal-burning power plant with few air pollution devices actually come in contact with during its lifetime?

23

Biological control basics

Predator-prey relationships & population dynamics

Studying population growth of a species helps us understand how it fits into our biosphere. If population growth were to be left completely unchecked, an organism could theoretically take over the world. This would result in what is called a "J" growth curve, since the population starts small (the lower portion of the "J") and then continues an upward trend.

Environmental factors, however, usually don't let this happen. Illnesses, predators, harsh weather, and many other factors control populations and keep them in check. This results in what is called an "S" growth curve. (See Fig. 23-1.) The top of the curve plateaus off because environmental factors control population growth.

Project overview

Aphids, also called plant lice, have remarkable reproductive capabilities. During most of their life cycle, they reproduce without mating and are born already pregnant. For most of the year, an aphid population is entirely composed of females, and they produce young parthenogenetically (without sex). Instead of laying eggs, like most

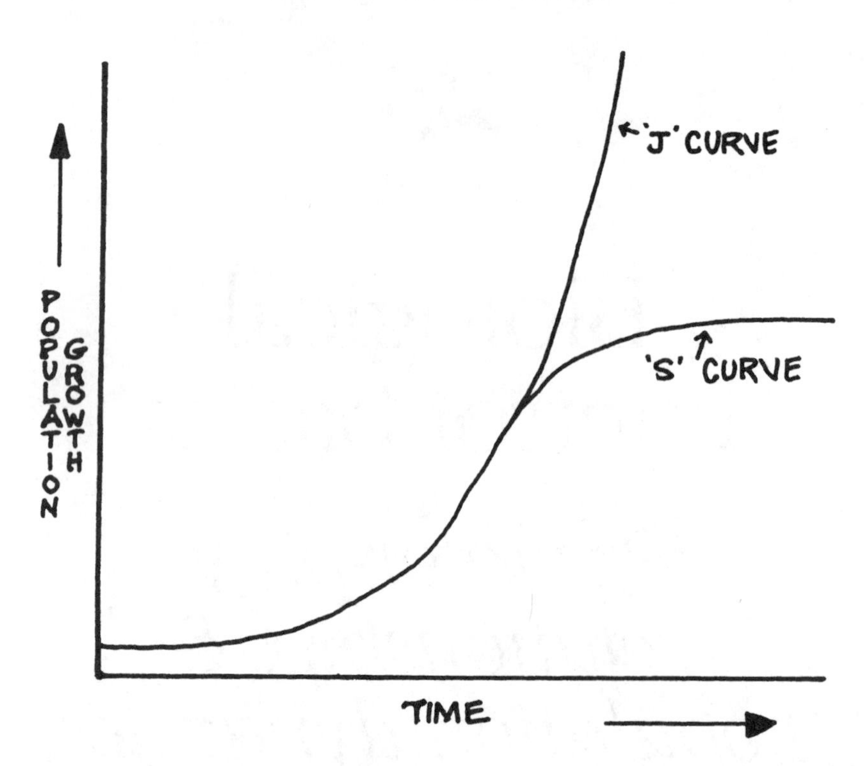

23-1 *Unchecked population growth produces a "J" curve while growth that meets with environmental resistance produces an "S" curve.*

insects, the female aphid produces live young, which are already pregnant with the next generation of offspring. The young aphids begin to feed immediately, and after developing into adults, they too are giving birth.

In the first part of this project, you will perform an experiment to see how many offspring a single female aphid can produce in her lifetime. In the second part, you will mathematically calculate a "doomsday chart" using the data you collected in the first part of the experiment. This chart will determine how long it takes for a single female aphid to generate a population of one million insects and beyond, assuming no mortality.

In the third part of this project you'll see how this population can be controlled. You'll figure in the impact of a natural predator, such as a lady beetle larvae, which devours large numbers of aphids. How does this predator control the population growth of aphids? How many offspring can a single aphid produce? How long would it take for one aphid to produce a population of one million?

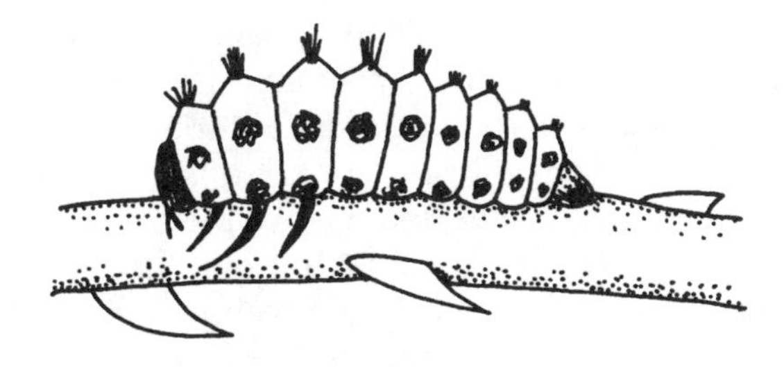

Materials list

- Plant infested with aphids (You can find one in a local field, garden, or possibly greenhouse, or you can create your own by placing a bean plant outdoors in a very sunny spot during the summer months in most parts of the country. It will be naturally infested with aphids within a few weeks.)
- Nylon material (such as pantyhose)
- String
- Forceps or fine tweezers
- Small paint brush (like those used to paint model airplanes)
- Magnifying glass
- Calculator
- Graph paper
- Electronic scale (optional)

Procedures

To begin the first part of this project, carefully examine the stem of the plant infested with aphids. Look for a small individual without wings. This is an immature nymph. To isolate this individual, use the brush or tweezers to push off all the other aphids nearby. You want this individual to be clear of other aphids in the colony by at least three inches in both directions on the stem.

Do not pick up the selected aphid while its beak is inserted into the plant. That would tear off its mouthparts and kill it. If necessary, you can gently pick up an aphid using the brush and place it on a stem without other aphids.

Once an individual is isolated, enclose that part of the plant stem (with the single aphid) in a fine nylon material such as pantyhose, so no other aphids can crawl or fly into that area. To do this, cut a nylon strip, four inches wide and wrap it around the stem, so it surrounds the plant loosely. Use string to tie each end of net to the plant stem. Make it tight enough so aphids can't crawl in or out, but don't crush or break the stem.

Create a similar setup on a different plant, or at least a different part of the same plant, so you'll have two running concurrently. Take notes about the date and time you begin the experiment. Return every two or three days, and count any new aphids (offspring) within the enclosures. To do this, gently untie and remove the nylon so you can get a clear view of the stem and the aphids. Count the new offspring and then remove them from the enclosure with the brush. (The offspring will be smaller than the mother.) Once again try not to crush them as they are removed. If the original aphid falls off the stem, carefully place it back on the stem using the brush. Replace the netting over the stem after each observation. Do this with each of the two plants being tested.

Continue your observations until offspring are no longer being produced. The original aphid will die shortly thereafter. Keep a record of the date and the number of offspring for each aphid. How many offspring did each aphid produce? How long did it take to produce the offspring?

With the information obtained in the first part of the project, you can now perform the second part. Calculate how long it would take to get one million aphids starting with a single female and assuming all her offspring survived. Graph this J-curve.

Analysis

How many offspring did a single aphid produce? How many offspring would theoretically be produced in future generations if they all survived? The results should represent a typical J-curve. How long would it take for the population to grow to one million individuals? How about ten million or one billion?

With few exceptions, morbidity and mortality factors keep populations from actually creating a J-curve. Environmental resistance, such as predators, disease, and toxic substances, result in a population growth called an S-curve. What types of organisms, if any, have populations that truly produce a J-curve?

Going further

After you calculate the numbers under ideal conditions, be more realistic. Vast numbers of insects don't survive to reproduce. Some are eaten by predators, some die of disease, and some will succumb to the environment. Incorporate into your figures assumptions about mortality. For example, assume that 30% of all the young are destroyed by predators, disease, harsh weather conditions, and do on. How does this impact the curve?

Suggested research

- Read about the growth curve of *Homo sapiens.*
- Read about the study of population dynamics.
- How does this project relate to the real world of biological control of insect pests? Why would this type of research be necessary before proceeding with a plan to release a beneficial insect in an effort to control a harmful one?

Using metrics

Most science fairs require all measurements be taken using the metric system rather than English units. Meters and grams, which are based on powers of ten, are actually far easier to use during your experimentation than feet and pounds.

You can convert English units into metric units if need be, but it is easier to simply begin with metric units. If you are using school equipment, such as flasks or cylinders, check the markings to see if any use metric units. If you are purchasing your glassware (or plastic ware), be sure to order metric markings.

Conversions from English units to metric units are given below, along with their abbreviations as used in this book. (All conversions are approximations.)

Length

one inch (in) = 2.54 centimeters (cm)
one foot (ft) = 30 centimeters (cm)
one yard (yd) = 0.90 meters (m)
one mile (mi) = 1.6 kilometers (km)

Volume

one teaspoon (tsp) = 5 milliliters (ml)
one tablespoon (tbsp) = 15 milliliters (ml)
one fluid ounce (fl oz) = 30 milliliters (ml)
one cup (c) = 0.24 liters (l)
one pint (pt) = 0.47 liters (l)

one quart (qt) = 0.95 liters (l)
one gallon (gal) = 3.80 liters (l)

Mass

one ounce (oz) = 28.00 grams (g)
one pound (lb) = 0.45 kilograms (kg)

Temperature

32° Fahrenheit (F) = 0° Celsius (C)
212° Fahrenheit (F) = 100° Celsius (C)
(See Fig. A-1.)

A-1
Use a thermometer like this to convert Celsius to Fahrenheit, and vice versa.

Scientific supply houses

You can order equipment, supplies and live specimens for projects in this book from these companies.

Blue Spruce Biological Supply Company
221 South St.
Castle Rock, CO 80104
(800) 621-8385

The Carolina Biological Supply Company
2700 York Rd.
Burlington, NC 27215
Eastern U.S.: (800) 334-5551
Western U.S.: (800) 547-1733

Connecticut Valley Biological
82 Valley Rd.
P.O. Box 326
Southampton, MA 01073

Fisher Scientific
4901 W. LeMoyne St.
Chicago, IL 60651
(800) 955-1177

Frey Scientific Company
905 Hickory Ln.
P.O. Box 8101
Mansfield, OH 44901
(800) 225-FREY

Nasco
901 Janesville Ave.
P.O. Box 901
Fort Atkinson, WI 53538
(800) 558-9595

Nebraska Scientific
3823 Leavenworth St.
Omaha, NB 68105
(800) 228-7117

Powell Laboratories Division
19355 McLoughlin Blvd.
Gladstone, OR 97027
(800) 547-1733

Sargent-Welch Scientific Company
P.O. Box 1026
Skokie, IL 60076

Southern Biological Supply Company
P.O. Box 368
McKenzie, TN 38201
(800) 748-8735

Ward's Natural Science Establishment, Inc.
5100 W. Henrietta Rd.
Rochester, NY 14692
(800) 962-2660

Ward's Natural Science Establishment, Inc.
815 Fiero Ln.
P.O. Box 5010
San Luis Obispo, CA 93403
(800) 872-7289

Glossary

abstract A brief, written overview describing your project, usually less than 250 words and often required at science fairs.

aerobic The presence of oxygen.

agar A gelatin-like substance that is made from seaweed. Used as a solid support for microbial cultures.

algae Single-celled organisms that contain chlorophyll and photosynthesize. Some live in large colonies and are macroscopic.

amoeba A common protozoan.

anaerobic The absence of oxygen.

autoclave A machine that sterilizes objects, such as glassware, by combining moist heat (steam) and pressure.

autotroph An organism that can produce its own food (by photosynthesis or from inorganic chemical energy). Also called a producer.

backboard The vertical, self-supporting panel used in your science fair display. The board usually displays the problem and hypothesis and photographs of the experimental set-up, organisms, and other important aspects of the project, as well as analyzed data in the form of charts and tables. Most fairs have size limitations for backboards.

bacteria A single-celled, microscopic organism that reproduces by fission and has no nuclear membrane.

biochemistry The study of biochemical reactions.

biodiversity The vast diversity of organisms on our planet and the importance of all.

biological control The use of beneficial organisms to control harmful (pest) organisms.

cell The basic unit of life. All cells are bags containing a liquid interior (the cytoplasm). The bag itself is the cell membrane.

cilia Tiny, short, hair-like structures used by some one-celled organisms to move.

ciliates Protozoans with cilia. *Paramecia* is an example of a ciliate.

colony A population of cells growing on a solid medium.

community All the populations living within a specified area.

computer modeling The use of computers to analyze existing data to make projections about what will happen in the future.

consumer An organism that must consume (eat) its food, as opposed to producers that make their food.

control group A test group that provides a baseline for comparison, where no experimental factors or stimulus are introduced.

culture (*noun*) Container of microbes with all the ingredients necessary for their survival. (*verb*) Growth of microbes.

decomposer Organism able to breakdown dead organic material (such as the dead bodies of animals or dead plant leaves). Fungi and many bacteria are decomposers.

dependent variable A variable that changes when the independent (also called experimental) variable changes. For example, if testing the mortality (death) rate of organisms living in soil exposed to pesticides, the mortality rate is the dependent variable, and the pesticides are the independent variable.

desiccation The loss of all water.

detritus Decomposing organic matter.

display Entire science fair exhibit, of which the backboard is a part.

ecosystem The living (organisms) and non-living (soil, water, etc.) components of a specified area, such as a pond or forest, and interactions that exist between all these components.

enzyme A protein that catalyzes (helps) a biochemical reaction to occur.

eukaryotic A type of organism whose cells have internal organelles and internal membranes, such as a nucleus. All non-bacterial organisms are eukaryotic, including the higher plants and animals.

exoenzyme An enzyme that is excreted outside of a cell.

exoskeleton The external supporting and protective structure of an arthropod such as an insect or lobster.

experimental group A test group that is subjected to experimental factors or stimulus for the sake of comparison with the control group. The experimental group is the one exposed to the factor being tested. For example, a plot of soil containing organisms exposed to varying amounts of pesticides.

experimental variable The aspect or factor to be changed for comparison during an experiment. For example, the amount of pesticide that soaks into the soil. Also called the independent variable.

flagellates Protozoans that use flagellum to move.

flagellum A tail-like structure used by microorganisms to move.

food chain A simple representation of "who eats what," represented by one-to-one relationships.

food web A representation of "who eats what" in an ecosystem, showing multiple feeding relationships. In other words, all the food chains linked together.

fungus Primitive organisms that feed on decaying organic matter. They play an important role in decomposition and recycling nutrients back into the environment.

habitat The place where an organism lives. For example, an aquatic or terrestrial habitat.

habituation The gradual reduction of a response to an event, such as a stimulus.

heterotroph Organisms that require an external source of organic chemical energy (food) to survive, as opposed to autotrophic. Same as a consumer.

host The organism which supports the life of a parasite. For example, a dog is a host for a tick.

hypothesis An educated guess, formulated after thorough research, to be shown true or false through experimentation.

indigenous Organisms that naturally live in an area, as opposed to foreign or exotic species that are introduced from elsewhere.

infection A growth of microorganisms within a host, causing illness in the host.

inorganic matter Substances that are not alive and did not come from decomposed organisms.

invertebrates Organisms with no backbones, such as insects, starfish, and lobsters.

journal Contains all notes on all aspects of a science fair project from start to finish. Also called the project notebook.

leaf litter Partially decomposed leaves, twigs, and other plant matter that have recently fallen to the ground, forming a ground cover.

macroscopic Something that is large enough to be seen with the unaided eye.

metabolism The sum of the physical and biochemical reactions necessary for life.

metamorphosis The change in body form during an insect's development.

microbe A small organism visible only with a microscope. Could be a bacteria, algae, fungi, protist, or virus.

morphology The study of the appearance of an organism, including its shape, texture, and color.

nematodes Small, unsegmented microscopic roundworms found in most habitats in great numbers. Most are harmless, but a few are parasitic.

nucleus A membrane-enclosed structure that contains genetic material in a eukaryotic cell.

observations A form of qualitative data collection.

organelle A membrane-enclosed structure within a cell in eukaryotic organisms.

organic Substances that compose living or dead, decaying organisms and their waste products. Carbon is the primary element.

parasite An organism that lives in or on one or more organisms (hosts) for a portion of its life. The host is usually not killed in the process.

parasitoid An animal that lives in another organism (host) and kills the host during its development.

parthenogenetic reproduction The ability to reproduce without a mate (reproduction without the fertilization of the egg).

pathogens Organisms that cause disease in other organisms.

populations All the members of the same species living in a specific area. For example, the population of silver foxes living in Maine.

population dynamics The study of populations and factors that affect them.

pheromone A chemical that communicates information between members of the same species.

predator An animal (consumer) that eats other animals for nourishment.

producer An organism that makes its own chemical energy (food), usually using energy from the sun.

Protista A kingdom of living things, composed of single-celled eukaryotes that do not have a cell wall. Some have chlorophyll, while others do not.

Protozoa A group of Protists that do not contain chlorophyll. Complex, single-celled animals (eukaryotes).

qualitative studies Experimentation where data collection involves observations but no numerical results.

quantitative studies Experimentation where data collection involves measurements and numerical results.

raw data Any data collected during the course of an experiment that has not been manipulated in any way. As opposed to smooth data.

research Locating and studying as much of the existing information about a subject as possible. Also called a literature search.

resolving power The smallest distance between two objects in which the two objects can still be distinguished from one another. If the two objects are beyond the resolving power of a microscope, the two objects appear as one.

scavenger An organism that consumes dead organic matter.

scientific method The basic methodology of all scientific experimentation: stating a problem to be solved or question to be answered, formulating a hypothesis, and performing experimentation to determine if the hypothesis is true or false. This includes data collection and analysis and arriving at a conclusion.

smooth data Raw data that has been manipulated to provide understandable information. Often presented in graphs and charts that represent totals, averages, and other numerical analysis.

species Organisms with the potential to breed and produce viable offspring.

statistics Analysis of numerical data.

sterile The absence of all life.

stimulus An event that prompts a reaction or a response.

survey collection A collection of organisms from a certain habitat or area.

variables A factor that is changed to test the hypothesis.

vertebrates Animals with backbones, such as reptiles, amphibians, birds, and mammals.

virus A package of genetic material surrounded by a protein capsule that requires a living host to reproduce.

Helpful books

The following field guides will help you identify organisms mentioned in this book:

Arnett, R. & R. Jacques. *Simon & Schuster's Guide to Insects.* Simon & Schuster: New York, NY, 1981

Bland, R.G. & H.E. Jaques. *How to Know the Insects.* W.C. Brown Co. Pub.: Dubuque, IA, 1978

Booth, Ernest S. *How to Know the Mammals.* W.C. Brown Co. Pub.: Dubuque, IA, 1970

Borror, D.J. & D.M. DeLong. *A Field Guide to the Insects.* Peterson Field Guide. Houghton Mifflin: Boston, MA, 1970

Borror, D.J. & R.E. White. *A Field Guide to the Insects of America North of Mexico.* Houghton Mifflin: Boston, MA, 1970

Burch. *How to Know Eastern Land Snails.* W.C. Brown Co. Pub.: Dubuque, IA, 1962

Chu, H.F. *How to Know the Immature Insects.* W.C. Brown Co. Pub.: Dubuque, IA, 1949

Dashefsky, H.S. & J.G. Stoffolano. *A Tutorial Guide to the Insect Orders.* Burgess Pub. Co.: Minneapolis, MN, 1977

Jahn, T.L., E.C. Bovee & F.F. Jahn. *How to Know the Protozoa, 2nd Ed.* W. C. Brown Co. Pub.: Dubuque, IA, 1979

Jaques, H.E. *How to Know Living Things.* W.C. Brown Co. Pub.: Dubuque, IA, 1946

Kaston. *How to Know Spiders.* W.C. Brown Co. Pub.: Dubuque, IA, 1952

Lehmkuhl, D.M. *How to Know the Aquatic Insects.* W.C. Brown Co. Pub.: Dubuque, IA, 1979

McCafferty, W.P. *Aquatic Entomology.* Jones & Bartlett: New York, NY, 1982

Schultz. *How to Know Marine Isopod Crustaceans.* W.C. Brown Co. Pub.: Dubuque, IA, 1969

If you are new to science fairs, here are some good books that cover all aspects of entering a science fair:

Bombaugh, Ruth. *Science Fair Success.* Enslow Publishers: Hillside, NJ, 1990

Irtz, Maxine. *Science Fair—Developing a Successful and Fun Project.* TAB/McGraw-Hill: Blue Ridge Summit, PA, 1987

Tocci, Salvatore. *How To Do A Science Fair Project.* Franklin Watts: New York, NY, 1986

The following books can all be used for additional science fair project ideas. Although not specifically about zoology, many involve animals or can be adapted to create zoological projects.

Barr, George. *Science Projects for Young People.* Dover Publications: New York, NY, 1964

Berman, William. *Exploring With Probe and Scalpel—How to Dissect—Special Projects for Advanced Studies.* Prentice Hall Press: New York, NY, 1986

Bochinski, Julianne. *The Complete Handbook of Science Fair Projects.* Wiley & Sons, Inc.: New York, NY, 1991

Bonnet, Robert & G. Daniel Keen. *Environmental Science: 49 Science Fair Projects.* TAB/McGraw-Hill, Inc.: Blue Ridge Summit, PA, 1990

Bybee, Dr. Rodger. *Acid Rain: Science Projects.* The Acid Rain Foundation, Inc.: St. Paul, MN, 1987

Dashefsky, H. Steven. *Insect Biology: 49 Science Fair Projects.* TAB/McGraw-Hill: Blue Ridge Summit, PA, 1993

Dashefsky, H. Steven. *Microbiology: 49 Science Fair Projects.* TAB/McGraw-Hill: Blue Ridge Summit, PA, 1994

Durant, Peggy Raife. *Prize Winning Science Fair Projects.* Scholastic, Inc.: New York, NY, 1991

Gardner, Robert. *More Ideas for Science Fair Projects.* Franklin Watts: New York, NY, 1989

Harlow, Rosie & Gareth Morgan. *175 Amazing Nature Experiments.* Random House: New York, NY, 1991

Irtz, Maxine. *Blue-Ribbon Science Fair Projects.* TAB/McGraw-Hill: Blue Ridge Summit, PA, 1991

Kneidel, Sally Stenhouse. *Creepy Crawlies and the Scientific Method.* Fulcrum Pub.: Golden, CO, 1993

Sheehan, Kathryn & Mary Waidner, Ph.D. *Earth Child: Games, Stories, Activities, Experiments & Ideas About Living Lightly on Planet Earth.* Council Oak Books: Tulsa, OK, 1991

Sisson, Edith A. *Nature with Children of All Ages.* Prentice-Hall Press: New York, NY, 1982

Tant, Carl. *Science Fair Spelled W-I-N.* Biotech Pub. Angleton, TX, 1992

VanCleave, Janice. *A+ Projects in Biology.* John Wiley & Sons, Inc.: New York, NY, 1993

VanCleave, Janice. *Biology for Every Kid: 101 Easy Experiments that Really Work.* John Wiley & Sons, Inc.: New York, NY, 1990

Witherspoon, James D. *From Field to Lab, 200 Life Science Experiments for the Amateur Biologist.* TAB/McGraw-Hill: Blue Ridge Summit, PA, 1993

Insects are often used as subjects in science fair projects. The following books provide more technical information about many of the insects used in these projects:

Imes, Rick. *The Practical Entomologist.* Simon & Schuster, Inc.: New York, NY, 1992 (for the hobbyist)

Romoser, William S. & John G. Stoffolano. *The Science of Entomology.* W. C. Brown Pub.: Dubuque, IA, 1993 (a college text)

For information about the International Science and Engineering Fairs and valuable information about Adult Sponsorship, contact

The Science Service
1719 N. Street N.W.
Washington, DC 20036
(202) 785-2255

Index

Illustrations are in **boldface**.

About the author

Steve Dashefsky is an adjunct professor of environmental science at Marymount College in Tarrytown, New York. He is the founder of the Center for Environmental Literacy, which was created to educate the public and business community about environmental topics. He holds a B.S. in biology and an M.S. in entomology and is the author of over ten books that simplify science and technology.